Atmospheric Impacts
of the Oil and Gas Industry

Atmospheric Impacts of the Oil and Gas Industry

Eduardo P. Olaguer
Houston Advanced Research Center
The Woodlands, TX, United States

ELSEVIER
AMSTERDAM • BOSTON • HEIDELBERG • LONDON
NEW YORK • OXFORD • PARIS • SAN DIEGO
SAN FRANCISCO • SINGAPORE • SYDNEY • TOKYO

Elsevier
Radarweg 29, PO Box 211, 1000 AE Amsterdam, Netherlands
The Boulevard, Langford Lane, Kidlington, Oxford OX5 1GB, United Kingdom
50 Hampshire Street, 5th Floor, Cambridge, MA 02139, United States

Notices
Knowledge and best practice in this field are constantly changing. As new research and experience broaden
our understanding, changes in research methods, professional practices, or medical treatment may become
necessary.

Practitioners and researchers must always rely on their own experience and knowledge in evaluating and
using any information, methods, compounds, or experiments described herein. In using such information or
methods they should be mindful of their own safety and the safety of others, including parties for whom they
have a professional responsibility.

To the fullest extent of the law, neither the Publisher nor the authors, contributors, or editors, assume any
liability for any injury and/or damage to persons or property as a matter of products liability, negligence or
otherwise, or from any use or operation of any methods, products, instructions, or ideas contained in the
material herein.

British Library Cataloguing-in-Publication Data
A catalogue record for this book is available from the British Library

Library of Congress Cataloging-in-Publication Data
A catalog record for this book is available from the Library of Congress

ISBN: 978-0-12-801883-5

For Information on all Elsevier publications
visit our website at https://www.elsevier.com

**Working together
to grow libraries in
developing countries**

www.elsevier.com • www.bookaid.org

Publisher: Candice Janco
Acquisition Editor: Louisa Hutchins
Editorial Project Manager: Emily Thomson
Production Project Manager: Vijayaraj Purushothaman
Designer: Matthew Limbert

Typeset by MPS Limited, Chennai, India

Dedication

This book is dedicated with heartfelt thanks
to my mentors over the years,
especially Ronald Prinn, Ka Kit Tung, Joseph Pinto,
Harvey Jeffries, and Charles Kolb.

Contents

8. Ambient Air Monitoring and Remote Sensing

9. Data Assimilation and Inverse Modeling

10. Photochemical Simulation

11.   MultiScale Impact Assessment

12.   Emission Controls

Introduction: Definition of the Problem

Energy drives growth. However, the mining of energy sources and the economic growth derived from it often come at the expense of degrading the environment in which we live. If this degradation were simply confined to isolated pockets, it would be easier to ignore for the sake of wealth creation and human progress. But this is not the case, for as science and technology become increasingly sophisticated, we perceive more and more that the Earth is a single contiguous and interdependent system. The atmosphere is historically the first of the Earth's environmental subsystems to command this realization, resulting in the widespread desire and search for a sustainable human civilization as expressed in global environmental treaties.

The purpose of this book is to provide the scientific and technological tools that can help realize the dream of sustainability, at least in part, by addressing an area of developing public concern, namely the explosion of unconventional oil and gas development made possible by the techniques of horizontal drilling and hydraulic fracturing, and the resulting emissions of atmospheric pollutants, sometimes in the midst of residential communities. In many places, this problem has resulted in an unfortunate polarization between the oil and gas industry and its customers, and even between the State and its own citizens. This is partly due to the lack of information on the environmental and human health risks to which the public has been exposed. In particular, we need answers to the following questions:

- How much of each particular chemical species of public concern is emitted by the oil and gas industry, and what is the distribution in time and space of the relevant emissions?
- How are the atmospheric pollutants emitted by the oil and gas industry chemically transformed and distributed throughout the environment?
- What is the exposure of human beings and sensitive ecological populations to these pollutants and their by-products of degradation?
- What are the hazards that attend human and ecological exposure to air pollution from the oil and gas industry, and can the risks associated with these hazards be quantified?
- What are the available methods to mitigate the risks attributable to atmospheric pollution by the oil and gas industry?

Although this book does not provide comprehensive answers to all these questions, it does provide an awareness of how such answers can be found based on science and technology that has yet to be widely disseminated among industry, the policy and regulatory communities, and the general public. These new methods should help ensure the viability of oil and gas exploration and production until cleaner and more sustainable energy resources are universally adopted.

The book is divided into two parts. Part I is a technical and descriptive summary of air quality and global change issues relevant to the oil and gas industry, designed to be accessible to a broad audience of scientists, engineers, and policymakers. Part II assumes a technical literacy at the level of an applied science or engineering graduate student, and concisely but rigorously summarizes state-of-the-art methods pertaining to the analysis and solution of the problems identified in Part I. Real world applications from field studies, policy-relevant modeling assessments, and regulatory decisions from multiple geographic regions are presented as an aid to understanding. If this book empowers readers to expedite a transition to a more sustainable future, then it will have been well worth the time and effort required to write it.

Part I

The Issues that Matter

Chapter 1

A Brief History of Oil and Gas Development From an Environmental Perspective

Chapter Outline

THE RISE OF UNCONVENTIONAL RESOURCES

Oil and natural gas, together known as petroleum, are fossil fuels derived from various types of sub-surface geological formations. Natural gas is composed principally of methane and other light hydrocarbons, while oil is made up of heavier alkanes, cycloalkanes, aromatics, and sulfur compounds. Crude oil is dark and viscous, while the term "condensate" refers to clear and volatile liquid petroleum.

Conventional oil and gas resources are those in which petroleum naturally flows through the source rock. Over the last few decades, however, oil and gas have been increasingly mined from unconventional reservoirs such as shale, tight sands, and coal beds, for which conventional drilling techniques are insufficient (see Fig. 1.1). The exploitation of unconventional resources has been made possible by horizontal drilling technology, which radically increases the sub-surface volume accessible via a single well pad, and more controversially by hydraulic fracturing, in which water, sand, and chemicals are used to break up dense rock formations deep underground, rendering them more porous. As a result of these technologies, the US Energy Information Administration (EIA) estimated that as of January 1, 2013 there were about 2276 trillion cubic feet (Tcf) of technically recoverable dry natural gas in the United States, enough natural gas to last about 84 years at the 2013 rate of consumption (EIA, 2015).

Atmospheric Impacts of the Oil and Gas Industry. DOI: http://dx.doi.org/10.1016/B978-0-12-801883-5.00001-2

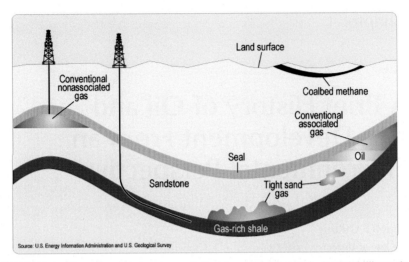

FIGURE 1.1   Illustration of various types of oil and gas resources and horizontal drilling technology. *Energy Information Administration (EIA), 2015. Annual energy outlook 2015 with projections to 2040. Office of Communications, Washington, DC.*

The exploitation of unconventional resources has resulted in a boom in US natural gas production since the turn of the century. The share of total gas production from unconventional reservoirs more than doubled (31−67%) from 2000 to 2011 (Moore et al., 2014). Shale gas production, in particular, is expected to increase by 73% from 11.3 Tcf in 2013 to 19.6 Tcf in 2040 (EIA, 2015).

It is the shale gas revolution, in particular, that has attracted the most environmental concern, largely because of the many and extensive shale plays in the United States (see Fig. 1.2), and because the mining of shale gas has penetrated even urban areas, beginning historically with major cities in the Barnett Shale of Texas, such as Fort Worth and Arlington. It was in the Barnett that George P. Mitchell, the founder of Mitchell Energy and Development Corporation, pioneered the extraction of shale gas using horizontal drilling and hydraulic fracturing techniques during the 1980s and 1990s (Wang et al., 2014). Since then, drilling and other exploration and production activities have proliferated throughout the Barnett, even amid residential neighborhoods, with required setbacks (distances between oil and gas sites and sensitive receptors such as dwellings) as little as a few hundred feet (Fry, 2013). This phenomenon has replicated itself to varying degrees in other shale plays throughout the country, drawing the ire of local citizens and activists in response to nuisances and alleged health hazards attending the mining of shale petroleum. Such conflicts have been exacerbated by split estate laws that separate the ownership of surface property and minerals, effectively denying landowners control over oil and gas activities.

Shale plays in the Lower 48 states

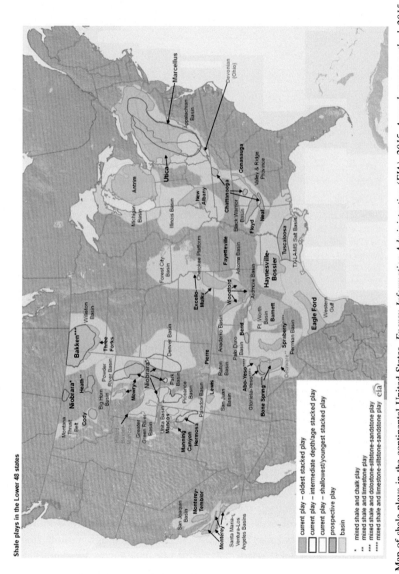

FIGURE 1.2 Map of shale plays in the continental United States. *Energy Information Administration (EIA), 2015. Annual energy outlook 2015 with projections to 2040. Office of Communications, Washington, DC.*

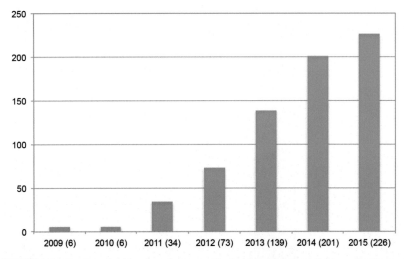

FIGURE 1.3   Number of publications that assess the impacts of shale or tight gas development per year, 2009—15. *Hays, J., Shonkoff, S.B.C., 2016. Toward an understanding of the environmental and publichealth impacts of unconventional natural gas development: a categorical assessment of the peer-reviewed scientific literature, 2009—2015. PLoS One 11, e0154164, http://dx.doi.org/10.1371/journal.pone.0154164.*

INVESTIGATIONS OF ENVIRONMENTAL AND HEALTH IMPACTS

The environmental and health impacts of unconventional oil and gas development have only recently become subjects of intense research. This is demonstrated by Fig. 1.3, reproduced from Hays and Shonkoff (2016), who surveyed the relevant peer-reviewed literature between 2009 and 2015. A literature review focusing exclusively on human health impacts was conducted by Werner et al. (2015), who ranked studies according to the strength of evidence on adverse environmental health outcomes, mainly due to air and water quality. They found only 109 relevant studies between 1995 and 2014, only 7 of which were considered highly relevant based on strength of evidence.

Among the factors limiting the validity of environmental health studies has been the protection of trade secrets, which prevents widespread knowledge of the identity and hazards of specific chemicals used in hydraulic fracturing. Colborn et al. (2011) used material safety data sheets to identify as many as 632 chemicals in shale gas operations. They found that 75% of the chemicals may negatively impact sensory, respiratory, or gastrointestinal systems, 50% may negatively affect the nervous, immune, and cardiovascular systems, 37% may disrupt the endocrine system, and 25% could cause cancer or genetic mutations.

Considerably more information is available in the scientific literature about environmental pollution due to shale petroleum mining than about corresponding human health impacts. The first major environmental concern to attract public attention was the potential contamination of water resources (Osborn et al., 2011). Such contamination may occur due to accidental spills and runoff of drilling fluids or produced water drawn from oil and gas reservoirs (flowback), the disposal of wastewater, the rupture of well cement or casings, or leaks from fractured rock. Rozell and Reaven (2012) concluded that disposal of drilling wastewater likely posed a much larger risk than other contamination pathways, while Vidic et al. (2013) argued that structural impairment of cement in wellbores was the most common mechanism for groundwater pollution due to oil and gas activities. In addition to concern about water quality, there is also concern about water quantity in areas where hydraulic fracturing occurs, since the process requires 2−4 million gallons of water per well (Wang et al., 2014). Injection or extraction of fluids during hydraulic fracturing or disposal of wastewater may also cause earthquakes (NRC, 2012).

Public attention to the atmospheric impacts of the oil and gas industry has focused mainly on the role of methane leaks in accelerating climate change. Natural gas, the dominant component of methane (CH_4), had originally been touted as a clean fuel, since its combustion leads to lower emissions of carbon dioxide (CO_2) relative to other fossil fuels such as coal. However, methane is itself a greenhouse gas (GHG). Methane's relative contribution to the radiative forcing of climate, i.e., its Global Warming Potential (GWP), is 28 times that of carbon dioxide over a 100-year time horizon, and 84 times that of carbon dioxide over a 20-year time horizon (IPCC, 2013).

Howarth et al. (2011) evaluated the GHG footprint of natural gas produced by hydraulic fracturing in shale formations, where:

$$GHG\ footprint = [CO_2\ emissions + (GWP \times CH_4\ emissions)]/(efficiency\ of\ use).$$

$$(1.1)$$

They found that 3.6−7.9% of methane from shale gas production escapes to the atmosphere over the lifetime of a well. Because of this, they concluded that the GHG footprint of shale gas is more than twice as great as that of coal on the 20-year horizon, and is comparable to that of coal when compared over 100 years. Howarth (2014) reconsidered available information in light of further research in the peer-reviewed literature and concluded that the original estimates of Howarth et al. (2011) were relatively robust. (More recent information on this issue is presented in Chapter 6, Greenhouse Gas Emissions and Climate Impacts.) Partly because of the issues raised by Howarth et al. (2011), the US Environmental Protection Agency (EPA) proposed regulations in 2015 requiring the capture and use of methane gas during completions, when wells are made ready for production.

As for local and regional air quality in the United States, much concern has been expressed regarding the oil and gas industry's possible contributions to urban nonattainment of the National Ambient Air Quality Standard (NAAQS) for ozone, a lung irritant that exacerbates asthma. This issue is most pronounced in the Barnett Shale encompassing much of the Dallas-Fort Worth (DFW) region, currently classified as in moderate nonattainment of the 2008 ozone NAAQS of 75 parts per billion (ppb) by volume averaged over 8 hours. The newer 2015 8-hour ozone standard is even more stringent at 70 ppb. Attainment of the federal ozone standard is reached when every regulatory monitor in the designated area achieves a design value at or lower than the standard, where the design value is defined as the 3-year average of the annual fourth highest daily 8-hour ozone maximum concentration at a monitoring site. As of 2015, the DFW ozone nonattainment area had 15 regulatory monitoring stations.

Public demonstrations of displeasure have occurred over the perceived weakness of the Texas State Implementation Plan (SIP) in promoting emission controls that are sufficient to bring the DFW area into ozone attainment, including more stringent measures to curb emissions of ozone precursors from oil and gas sites (Prince, 2016). As of this writing, the controversy has possibly opened the door to a rarely invoked measure in which the federal government steps in and creates its own Federal Implementation Plan (FIP), effectively overriding the State of Texas in regulating emission sources of Volatile Organic Compounds (VOCs) and nitrogen oxides ($NO_x = NO + NO_2$), the two main classes of ozone precursors.

SIP ozone attainment demonstrations are usually conducted with regional air quality models at horizontal resolutions of 4 km in the urban core, with coarser resolution elsewhere. These numerical models are used to assess the effectiveness of proposed strategies to control emissions of NO_x and VOCs from both stationary and mobile sources. Emission control strategies typically reduce ozone design values by only a few ppb. Kemball-Cook et al. (2010) conducted regional photochemical modeling to assess the ozone impact of petroleum mining in the Haynesville Shale at the Texas-Louisiana border. They found that projected 2012 ozone design values in northeast Texas and northwest Louisiana increased by up to 5 ppb due to oil and gas development in the Haynesville Shale. Olaguer (2012) deployed a micro-scale air quality model with finer horizontal resolution (200 m) than in SIP models. He showed that banks of compressor engines used at gas processing plants, as well as flares that burn natural gas, may be responsible for producing a few ppb of ozone at relatively short distances (2−10 km) downwind of oil and gas facilities, making their true impacts hard to detect by a sparse observational network of regulatory monitors and impossible to simulate with regional models.

Among the VOCs that are products of incomplete combustion of natural gas is the oxygenated hydrocarbon, formaldehyde (HCHO). This compound is especially important because it is both a precursor of atmospheric radicals that accelerate the formation of ozone, and a toxic air pollutant that may cause respiratory symptoms and cancer. Measurements of HCHO immediately next to oil and gas sites in the Barnett Shale have been known to exceed 100 ppb averaged over 1 hour (BSEEC, 2010; Olaguer 2012; Olaguer et al., 2016), well over the short-term (1 hour) Effects Screening Level (ESL) of 12 ppb used by the Texas Commission on Environmental Quality (TCEQ) as a guide to delineate acceptable human health risk.

Formaldehyde is one of 187 Hazardous Air Pollutants (HAPs) listed by the United States Clean Air Act, all of which have serious adverse human health or ecological impacts. Other HAPs include the aromatic hydrocarbons: benzene, toluene, ethyl benzene, and xylenes (BTEX), which are not only products of incomplete combustion of natural gas, but also fugitive emissions from petroleum storage and transfer. Benzene, in particular, is a known carcinogen. Olaguer et al. (2016) conducted advanced real time measurements of BTEX compounds at a natural gas production facility in the Eagle Ford Shale of South Texas. They found ambient concentrations of benzene of more than 200 ppb in the immediate vicinity ($\sim$ 100 m downwind) of a flare. By comparison, the TCEQ's short-term ESL for benzene is 54 ppb.

Assessments of the atmospheric impacts of the oil and gas industry are ultimately limited by poor knowledge of the emissions attributable to petroleum mining. Emission inventories used in regulatory applications are traditionally reported by industry based on standard estimation protocols, such as those supplied by the USEPA as part of the AP-42 database (EPA, 1991). These so-called "bottom-up" estimates are increasingly evaluated for accuracy using "top-down" methods, in which actual measurements of ambient pollutant concentrations are attributed to an emission source based on either statistical techniques or atmospheric models. For example, based on ground and aerial sampling of methane and an atmospheric transport model, Miller et al. (2013) found that methane emissions due to oil and gas extraction in the United States are underestimated by official emission inventories by roughly a factor of two or more.

The information presented in this section has been but a brief survey of environmental issues associated with oil and gas development. Those specifically pertaining to the atmosphere are the subject of the rest of this book, and will be discussed in considerably greater detail.

REFERENCES

Barnett Shale Energy Education Council (BSEEC), 2010. Ambient air quality study: natural gas sites, cities of Fort Worth and Arlington, Texas. Titan Engineering final report, Fort Worth, TX.

Colborn, T., Kwiatkowski, C., Schultz, K., Bachran, M., 2011. Natural gas operations from a public health perspective. Hum. Ecol. Risk Assess. Int. J. 17, 1039–1056.

Environmental Protection Agency (EPA), 1991. AP-42: Compilation of air pollutant emission factors. Available at: <http://www.epa.gov/ttnchie1/ap42> (accessed 03.02.16.).

Energy Information Administration (EIA), 2015. Annual energy outlook 2015 with projections to 2040. Office of Communications, Washington, DC.

Fry, M., 2013. Urban gas drilling and distance ordinances in the Texas Barnett Shale. Energy Policy 62, 79–89.

Hays, J., Shonkoff, S.B.C., 2016. Toward an understanding of the environmental and public-health impacts of unconventional natural gas development: a categorical assessment of the peer-reviewed scientific literature, 2009–2015. PLoS One 11, e0154164. Available from: <http://dx.doi.org/10.1371/journal.pone.0154164>.

Howarth, R.W., 2014. A bridge to nowhere: methane emissions and the greenhouse gas footprint of natural gas. Energy Sci. Eng. 2, 47–60.

Howarth, R.W., Santoro, R., Ingraffea, A., 2011. Methane and the greenhouse-gas footprint of natural gas from shale formations. Clim. Change 106. Available from: <http://dx.doi.org/10.1007/s10584-011-0061-5>.

Intergovernmental Panel on Climate Change (IPCC), 2013. Climate Change 2013: The Physical Science Basis. Contribution of Working Group I to the Fifth Assessment Report of the Intergovernmental Panel on Climate. Cambridge University Press, Cambridge and New York, NY.

Kemball-Cook, S., Bar-Ilan, A., Grant, J., Parker, L., Jung, J., Santamaria, W., et al., 2010. Ozone impacts of natural gas development in the Haynesville Shale. Environ. Sci. Technol. 44, 9357–9363.

Miller, S.M., Wofsy, S.C., Michalak, A.M., Kort, E.A., Andrews, A.E., Biraud, S.C., et al., 2013. Anthropogenic emissions of methane in the United States. Proc. Natl. Acad. Sci. USA. 110, 20018–20022.

Moore, C.W., Zielinska, B., Petron, G., Jackson, R.B., 2014. Air impacts of increased natural gas acquisition, processing and use: a critical review. Environ. Sci. Technol. 48, 8349–8359.

National Research Council (NRC), 2012. Induced Seismicity Potential in Energy Technologies, Washington, DC.

Olaguer, E.P., 2012. The potential near source ozone impacts of upstream oil and gas industry emissions. J. Air Waste Manag. Assoc. 62, 966–977.

Olaguer, E.P., Erickson, M.H., Wijesinghe, A., Neish, B.S., Williams, J., Colvin, J., 2016. Updated methods for assessing the impacts of nearby gas drilling and production on neighborhood air quality and human health. J. Air Waste Manag. Assoc. 66, 173–183.

Osborn, S.G., Vengosh, A., Warner, N.R., Jackson, R.B., 2011. Methane contamination of drinking water accompanying gas-well drilling and hydraulic fracturing. Proc. Natl. Acad. Sci. U.S.A. 108, 8172–8176.

Prince, J., 2016. Big Brown, Big Bad. Fort Worth Weekly, February 3.

Rozell, D.J., Reaven, S.J., 2012. Water pollution risk associated with natural gas extraction from the Marcellus Shale. Risk Anal. 32, 1382–1393.

Vidic, R.D., Brantley, S.L., Vandenbossche, J.M., Yoxtheimer, D., Abad, J.D., 2013. Impact of shale gas development on regional water quality. Science 340 (6134). Available from: <http://dx.doi.org/10.1126/science.1235009>.

Wang, Q., Chen, X., Awadhesh, N.J., Rogers, H., 2014. Natural gas from shale formation— the evolution, evidences and challenges of shale gas revolution in United States. Renewable Sustainable Energy Rev. 30, 1–28.

Werner, A.K., Vink, S., Watt, K., Jagals, P., 2015. Environmental health impacts of unconventional natural gas development: a review of the current strength of evidence. Sci. Total Environ. 505, 1127–1141.

Chapter 2

Overview of Oil and Gas Processes and Their Emissions to Air

Chapter Outline

EMISSION SOURCE CATEGORIES

Oil and gas fields are drilled when potential sources of petroleum have been identified through geophysical (e.g., gravitational, magnetic, seismic), remote sensing, or other techniques. After the wells have been drilled and completed, the production of crude oil and/or natural gas follows. Natural gas is first gathered from the field and processed at a gas plant before it is made available for consumer use, while crude oil is delivered to refineries. Petroleum products are transported away from the field over land via road, rail or pipeline networks, and over waterways by barges and tankers. The first phase of finding, drilling, and producing petroleum is referred to as the "upstream" or exploration and production sector, while the "midstream" sector refers primarily to the gathering, processing, storage, and transport of natural gas and crude oil. Finally, the "downstream" sector involves the refining and distribution of oil-based products. Here we are concerned mostly with the upstream and midstream sectors of the oil and gas industry.

Emissions to air occur throughout the entire chain of upstream and midstream processes. Major categories of emission activities include: (1) exploration and production; (2) gathering and processing; (3) combustion; (4) storage and transmission; and (5) waste disposal. We will discuss each of these emission source categories in turn.

Atmospheric Impacts of the Oil and Gas Industry. DOI: http://dx.doi.org/10.1016/B978-0-12-801883-5.00002-4

EXPLORATION AND PRODUCTION

Drilling operations are an important source of short-term air pollution. Well site preparation produces emissions from trucks and heavy machinery, such as dust and diesel particulate matter, and CO, NO_x, and VOCs from internal combustion engines. Hydraulic fracturing or "fracking" produces similar emissions, with additional pollution associated with VOCs and HAPs from fracking fluids, flowback, and produced water, as well as hazardous silica dust (used as "proppant" in shale rock along with manmade ceramic beads). Hydrogen sulfide (H_2S), which is flammable, asphyxiating, and lethal at high concentrations, is another pollutant that can be released by drilling activities, including emissions from flowback and produced water.

A typical drill rig is shown in Fig. 2.1. During drilling, well gases and broken bits of rock called "cuttings" may become entrained in a special lubricating mud designed to protect the drill bit and the walls of the wellbore. The drilling mud is degassed and sifted to remove cuttings before being stored in tanks for recycling. Separated gases are vented to the atmosphere or flared, while cuttings are diverted to waste pits prior to disposal. Pollutants generated by mud degassing include methane, H_2S, VOCs, and HAPs.

When a well reaches completion (i.e., is ready for production), an outer casing is installed in the well hole and heavy salt water used to prevent premature flow of petroleum. The potential flow of oil or gas is measured, and the gas that is produced during this testing is either flared or vented to the atmosphere, while extracted crude oil is stored in tanks. Well completion processes can lead to emissions of VOCs, HAPs, and methane, especially during "blowouts," when the well formation pressure significantly exceeds the completion fluid pressure.

At the end of completion, a combination of high-pressure valves known as a "Christmas tree" is installed to control oil and gas flow (see Fig. 2.2). Production from older wells may also be enhanced by "pump jacks" that artificially lift petroleum to the surface (see Fig. 2.3). Fugitive emissions of

FIGURE 2.1   Drill rig. *US Department of Energy.*

FIGURE 2.2 Christmas tree. *Wikipedia.*

FIGURE 2.3 Pump jack. *Bureau of Land Management.*

methane, VOCs, and HAPs may occur during the production phase due to leaks from valves and fittings or during process upsets, such as releases due to overpressure. Work-over and maintenance activities can also produce emissions from trucks and heavy machinery.

In mature gas wells, insufficient reservoir pressure may result in the accumulation of liquids, which may require removal to maintain production. This process, known as liquids unloading, is a potentially significant source of VOC and methane emissions (EPA, 2014).

GATHERING AND PROCESSING

Oil and natural gas must first undergo a number of processing steps prior to transport and consumption. Some preliminary treatment of wellhead gas may be performed as the gas is gathered from the field and compressed through pipelines that feed processing facilities, so that the line between gathering and processing is somewhat blurred. In this section, we provide a general overview of the various treatment steps required to process raw petroleum. The reader is referred to EPA (1999) and references therein for a more detailed discussion.

Gas and Liquid Separation

Gases and liquids extracted from a well must first be separated in a series of processes. Vertical, spherical, or horizontal separators are typically employed to separate natural gas from liquid hydrocarbons and water (see Fig. 2.4). Additional water and gas separation from oil emulsion streams may be required. Oil emulsions are broken using heat in heater treaters or electric energy in electrostatic coalescers. Water recovered during emulsion breaking is skimmed for remaining oil, filtered, and then stored in water tanks prior to disposal. The emissions from fuel combustion in heater treaters and from water storage and disposal are discussed further below.

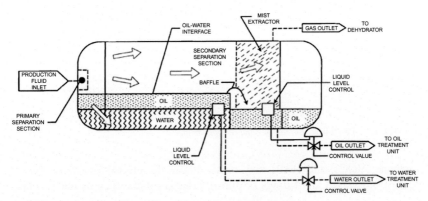

FIGURE 2.4 Three-phase separator. *US Department of Energy.*

Glycol Dehydration

Water must be removed from natural gas streams to prevent pipeline corrosion and the formation of hydrates. Tetraethylene, triethylene, diethylene, or ethylene glycol is added to the natural gas stream to absorb liquid water and water vapor. BTEX as well as other hydrocarbon VOCs and HAPs are likewise absorbed into the glycol stream. "Rich glycol" is glycol saturated with water that is sent to a still to remove water and hydrocarbons. "Lean glycol" suitable for reuse results from this regeneration process. A typical glycol unit is illustrated in Fig. 2.5.

Emissions from glycol dehydration include gases vented from the dehydrator flash tank or from the glycol regenerator. A gas-fired reboiler used in a glycol regenerator can be used to thermally oxidize hydrocarbons emitted from the regenerator process vent, although the reboiler can itself be a source of HAP emissions and fuel combustion gases. Gas-driven pumps may produce similar emissions, while fugitive releases may come from valves and fittings.

Methanol Injection

Methanol is often added as antifreeze or anticoagulant to natural gas streams via a gas-powered injection pump. Sources of pollution from this process include fugitive emissions from transfer line fittings and methanol storage tanks, and emissions from the injection pump itself.

Particulate Filtering

Solid impurities are removed from natural gas through a filter. Maintenance, repair, cleaning, and disposal of filter cartridges may result in emissions due to volatilization of hydrocarbons. Fugitive emissions from valves and flanges may also occur.

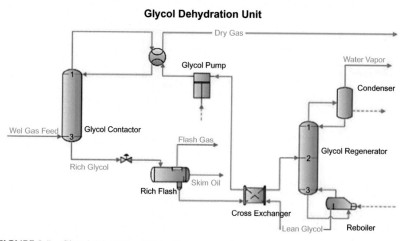

FIGURE 2.5 Glycol dehydrator. *Wikipedia.*

Gas Sweetening and Sulfur Recovery

Raw natural gas is "sweetened" by removing corrosive acids such as H_2S and CO_2. Acids are scavenged using amines, glycol ether solvents, fixed sorption beds, or other means. A sulfur recovery process may then be applied to exhaust gases. Natural gas sweetening processes are potential sources of VOC, HAP, H_2S, SO_2, NO_x, and CO emissions. For example, amine regenerators used to heat amine solution release acid gases through the still vent (see Fig. 2.6). Fixed sorption bed processes can vent sour (unsweetened) gas from flash tanks, emit exhaust from process heaters or release fugitives. Emission sources during sulfur recovery include process vent streams, leaks from valves, flanges and seals, and combustion exhaust.

Hydrocarbon Recovery

Hydrocarbons are often separated and recovered from natural gas streams using a variety of processes, including cryogenic expansion, refrigeration, absorption, and adsorption (Gas Research Institute, 1994). Emission sources

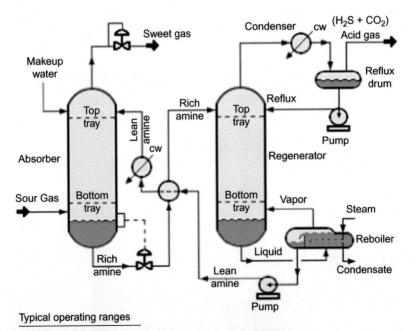

Typical operating ranges

Absorber : 35 to 50 °C and 5 to 205 atm of absolute pressure
Regenerator : 115 to 126 °C and 1.4 to 1.7 atm of absolute pressure
          at tower bottom

FIGURE 2.6   Amine unit. *Wikipedia.*

include exhaust from compressors and regenerators, flue gas from reboilers, fugitive emissions, and emissions due to maintenance activities.

Blowdown

Unexpected process upsets, as well as scheduled maintenance, startup, and shutdown of oil and gas facilities can lead to large transient releases known as emission events. For example, during so-called "blowdown" events, compressors are shutdown, and any remaining gas in equipment and pipelines must be vented or flared to reduce pressure prior to servicing and startup.

COMBUSTION

Combustion emissions of NO_x, CO, VOC, HAPs, H_2S, SO_2, methane, CO_2, and particulate matter (PM) result from a variety of oil and gas processes. Boilers and heaters provide heat and/or steam for electric power generation and for process units such as glycol and amine reboilers or oil and gas separators (i.e., heater treaters), or to maintain temperature within pipes, connections, or storage tanks. Stationary reciprocal internal combustion engines (RICE) and turbines have numerous purposes, including natural gas and refrigerant compression, generation of electricity, or the operation of equipment such as pumps. For example, compressor engines rated at thousands of horsepower are used to push natural gas through pipeline networks. A typical gas compressor is shown in Fig. 2.7.

Among the most notable combustion processes at oil and gas sites is flaring (see Fig. 2.8). Flares are control devices used to limit VOC and other emissions from storage tanks, glycol dehydrators, amine and sulfur recovery

FIGURE 2.7 Gas compressor. *US Department of Energy.*

FIGURE 2.8 Offshore oil platform flare. *Wikipedia.*

units, loading operations, and vent collection systems. They are also used to convert H_2S and other reduced sulfur compounds to SO_2. In the Bakken fields of North Dakota where crude oil is the economically preferred output to natural gas, an enormous amount of gas is flared due to the lack of infrastructure to process and transport the gas. At night, the resulting lights can be seen from space by satellites, and are as bright as the night lights from major urban areas in the United States (see Fig. 2.9).

STORAGE AND TRANSMISSION

A number of liquids used or generated during the production of oil and gas must be stored in tanks prior to delivery, use, or disposal (see Fig. 2.10). These liquids include crude oil, condensate, liquefied natural gas (LNG), produced water or brine, diesel fuel, and production chemicals such as methanol. Fugitive losses from storage tanks include working, breathing, and flash emissions. Working losses occur when liquids already in the tank are agitated during tank filling and emptying, while breathing or standing losses are due to vapor expansion of tank fluid accompanying normal changes in atmospheric temperature and pressure. Flashing occurs when liquids are exposed to rises in temperature or decreases in pressure while being transferred to storage tanks from production vessels.

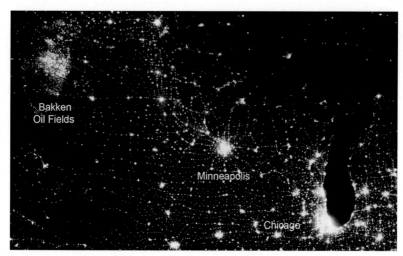

FIGURE 2.9   Satellite night lights. *US Department of Energy.*

FIGURE 2.10   Storage tanks. *Argonne National Lab.*

Emissions during petroleum transport and transmission occur due to loading losses and pipeline leaks. Emissions can also occur during pipeline pigging operations, when a physical device known as a pig (typically made of plastic, foam, or rubber) is forced through a pipeline by a compressed

FIGURE 2.11   Pipeline cleaning pig. *Wikipedia.*

gas such as nitrogen to assist in product transfer or separation, or in cleaning and maintenance activities (see Fig. 2.11). Residual vapors pushed through the pipeline by the pig may escape through vents or other openings.

WASTE DISPOSAL

Waste pits and other units used to treat, store, and transfer wastewater may emit VOCs and HAPs as well as methane and H_2S, if open to the atmosphere (see Fig. 2.12). Pits used to store hydrocarbon-laden cuttings may also be a source of VOC and HAP emissions.

Landfarming is a method of treatment and disposal in which drilling mud and cuttings are spread over and mixed into soils to reduce organic pollutants and attenuate metals (McFarland et al., 2009). Physical, chemical, and biological processes are used to control waste migration in the soil. Landfarming may result in emissions to air due to volatilization of chemical wastes.

OIL AND GAS SUPPLY CHAINS

Emissions from the oil and gas industry are distributed along supply chains, sequentially organized operations through which petroleum flows from production sites to the consumer. The oil and gas industry is characterized by a high degree of vertical integration, wherein firms that buy and sell from each other within the same supply chain are often joined together to maximize efficiency and minimize uncertainty. Separate supply chains exist for natural gas and for oil, as illustrated in Figs. 2.13 and 2.14, which track the flows of energy through various components of the industry. (Note that some natural gas is either lost or consumed as fuel along the natural gas supply chain.)

FIGURE 2.12 Waste pit. *Argonne National Lab.*

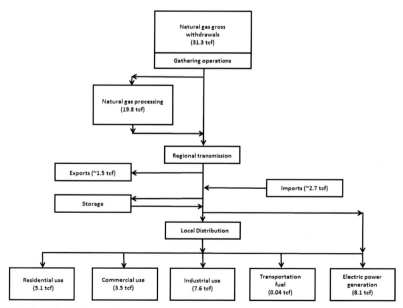

FIGURE 2.13 US natural gas supply chain, with 2014 flows in trillion standard cubic feet (tcf). *Allen, D.T., 2016. Emissions from oil and gas operations in the United States and their air quality implications. J. Air Waste Manage. Assoc. 66, 549–575.*

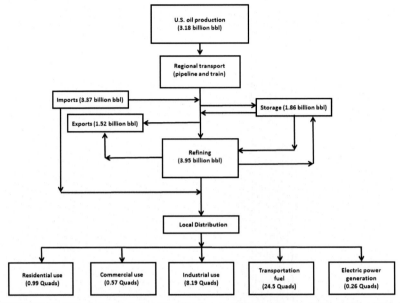

FIGURE 2.14   US oil supply chain, with 2014 flows in billions of barrels (billion bbl) and Quads (10^{15} BTU). Note that the number of Quads is only an approximation due to the difference in energy content of various oil products. *Allen, D.T., 2016. Emissions from oil and gas operations in the United States and their air quality implications. J. Air Waste Manage. Assoc. 66, 549–575.*

In 2011, the total emissions to air resulting from US oil and gas supply chains amounted to roughly 2.8 million tons of VOC, 14.5 million tons of NO_x and 226.4 million metric tons of greenhouse gases, expressed in CO_2 equivalents (Allen, 2016). In succeeding chapters, we will examine the environmental and human health impacts of these emissions.

REFERENCES

Allen, D.T., 2016. Emissions from oil and gas operations in the United States and their air quality implications. J. Air Waste Manage. Assoc. 66, 549–575.

Environmental Protection Agency (EPA), 1999. Preferred and alternative methods for estimating air emissions from oil and gas field production and processing operations. In: EIIP Technical Report Series, Volume 2: Point Sources. Available at: <https://www.epa.gov/sites/production/files/2015-08/documents/ii10.pdf>. Accessed 6/3/2016.

Environmental Protection Agency (EPA), 2014. Oil and natural gas sector liquids unloading processes. Office of Air Quality Planning and Standards. Available at: <https://www3.epa.gov/airquality/oilandgas/2014papers/20140415liquids.pdf>. Accessed 6/25/2016.

Gas Research Institute (GRI), 1994. Preliminary assessment of air toxic emissions in the natural gas industry, Phase I, topical report. GRI-94/0268. Chicago, IL.

McFarland M.L., Feagley S.E., Provin T.L., 2009. Land application of drilling fluids: Landowner considerations. Texas A&M AgriLife Extension Service. Available at: <http://soiltesting.tamu.edu/publications/SCS-2009-08.pdf>. Accessed 6/7/2016.

Chapter 3

Toxic Air Pollution on Neighborhood Scales

Chapter Outline

AN INVISIBLE THREAT

On October 23, 2015, the Southern California Gas Company (SoCal Gas) detected a leak in its Aliso Canyon storage facility. The site of the facility in the Santa Susana mountains on the outskirts of Los Angeles had once been an oil field owned by John Paul Getty, and was eventually completely drained of oil. The resulting cavity, which lies between 7100 and 9400 feet below ground, now stores natural gas up to a capacity of 84 billion standard cubic feet (scf), one of the largest reservoirs of its kind in the United States (Rich, 2016). After the initial detection on October 23, 2015, the Aliso Canyon blowout released over 97,000 metric tons of methane (Conley et al., 2016), the worst natural gas leak in US history.

In the affluent neighborhood of Porter Ranch, at the southern end of the Santa Susana mountains, residents found the smell of gas from the Aliso Canyon leak overwhelming, and complained of severe nosebleeds, respiratory problems, headaches, nausea, and vomiting. Rather than methane, which is odorless, the residents smelled sulfur compounds called mercaptans that are added to natural gas so that pipeline leaks can be more easily detected. While mercaptans are not known to cause long-term health problems, they may indicate the presence of more toxic gases, in particular those associated with the original crude oil from the reservoir. Shortly after the leak was detected, SoCal Gas measured short-term atmospheric mixing ratios of benzene as high as 30 ppb near the Aliso Canyon facility (Rich, 2016). Acute exposure to benzene can cause drowsiness, headaches, and eye, skin, and respiratory tract infections, while chronic exposure to even a few ppb of benzene can cause

Atmospheric Impacts of the Oil and Gas Industry. DOI: http://dx.doi.org/10.1016/B978-0-12-801883-5.00003-6
23

reproductive effects and blood disorders, including anemia and leukemia (WHO, 2000; EPA, 2003; Werner et al., 2015). As a safety measure, SoCal Gas temporarily relocated about 11,296 people as of January 7, 2016, including two schools (Maddaus, 2015).

The Aliso Canyon gas leak lasted for four months and in the process contributed as much greenhouse gas pollution as an entire third world country. The immediate impact of the blowout, however, was not due to its global warming repercussions but to its potential effects on human health and welfare at the neighborhood scale. These effects are far more palpable, yet still invisible from the standpoint of scientific evidence, more so perhaps than for climate change.

It does not take an obvious gas leak such as the Aliso Canyon blowout to create problematic exposures to toxic air pollutants. An example of this in the midstream and downstream oil and gas sectors was provided during the Benzene and other Toxics Exposure (BEE-TEX) field study in the vicinity of the Houston Ship Channel (Olaguer, 2015). During BEE-TEX, mobile real-time measurements of BTEX compounds (1 second time response, sub-ppb detection limits) were made using a chemical ionization technique known as Proton Transfer Reaction—Mass Spectrometry (PTR-MS) in the city of Galena Park, Texas. The city is the site of marine port terminals along the Ship Channel, several above-ground storage tank farms, and a network of underground pipelines carrying natural gas, crude oil, benzene, and refined products to and from nearby petrochemical facilities. The area had been placed by the Texas Commission on Environmental Quality (TCEQ) on a watch list for benzene, which sparked vigorous efforts by industry to detect fugitive leaks and reduce benzene emissions. These efforts included using helicopter-mounted infrared cameras to detect pollution plumes from short-term emission events. While these efforts were generally successful in reducing long-term benzene readings at local regulatory monitors, the mobile measurements during BEE-TEX revealed short-term spikes in ambient benzene concentration up to about 40 ppb (see Fig. 3.1).

An analysis of the mobile measurements in Galena Park by Olaguer et al. (2016) revealed that the largest benzene plumes were likely caused by invisible pipeline emissions below the detection limit of infrared cameras. Moreover, the pipeline leaks were apparently correlated with marine loading and unloading operations at a nearby port terminal, as might be expected if the pipelines were tapped due to product transfers at the docks (e.g., to refill depleted storage tanks). A disconcerting aspect of the BEE-TEX study results was that the pipeline emissions occurred in a residential neighborhood nestled closely beside port and industrial facilities. Such "fence line communities" bear the brunt of environmental injustice, all the more so when, unlike Porter Ranch in California, they are populated by socially and economically disadvantaged racial minorities, as is the case with Galena Park.

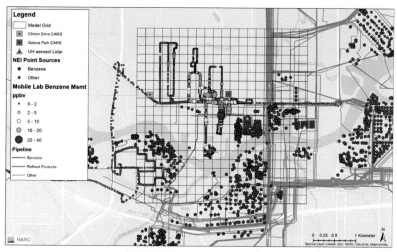

FIGURE 3.1   Map showing BEE-TEX mobile lab measurements of benzene on February 19, 2015, along with point sources (red pentagons for benzene, blue for other compounds) and pipeline network segments. The locations of Continuous Ambient Monitoring Stations (CAMS) are also indicated. *Olaguer, E.P., Erickson, M.H., Wijesinghe, A., Neish, B.S., 2016. Source attribution and quantification of benzene event emissions in a Houston Ship Channel community based on real time mobile monitoring of ambient air. J. Air Waste Manage. Assoc. 66, 164–172.*

THE MEASUREMENT DILEMMA

The use of advanced real-time monitoring techniques during the BEE-TEX field study was an attempt to overcome a major obstacle in the characterization of air toxics exposures at the neighborhood scale, namely the availability of measurement tools of sufficient power and scope to analyze pollution plumes that are highly variable in time and space. When confronted with the possibility of serious problems involving HAPs, regulators have typically resorted to conventional analytical methods. These include grab sampling techniques involving Summa canisters and 2,4-dinitrophenylhydrazine (DNPH) cartridges, which require off-line laboratory analysis (risking degradation of the air sample), and invariably suffer from poor temporal and spatial coverage. Cheap, fast, and portable photo-ionization detectors that measure total hydrocarbons can alleviate this problem to some degree, but cannot provide speciation information unless they are used to trigger the collection of whole air samples for later analysis.

At the high end of conventional measurement techniques are expensive monitoring stations equipped with a variety of automated sampling equipment (see Fig. 3.2), including gas chromatographs with flame ionization detection (GC-FID) for the measurement of speciated hydrocarbons. The equipment at a typical regulatory monitoring station is described in greater detail in Appendix A. Stationary monitors cannot, by their very nature, adjust to variable wind

FIGURE 3.2   Continuous Ambient Monitoring Station in DISH, Texas. *TCEQ.*

direction, so that a very dense network of such monitors is required to reliably measure sporadic large plumes emanating from oil and gas activities, many of which have been detected without quantification by environmental activists using infrared cameras.

The dependence on a relatively sparse regulatory monitoring network is the major flaw behind the study of Bunch et al. (2014), who analyzed 4.6 million VOC measurements from seven automated gas chromatograph (auto-GC) stations established by the TCEQ in the Barnett Shale. Bunch et al. (2014) concluded on the basis of these measurements that shale production activities did not result in community-wide ambient air exposures that would pose any health concerns. However, the auto-GC stations that produced the measurements are better suited to the characterization of regional air pollution than the detection of neighborhood scale plumes.

ALLEGED HEALTH EFFECTS OF OIL AND GAS SITE EMISSIONS

Unlike Bunch et al. (2014), McKenzie et al. (2012) analyzed air samples in Garfield County, Colorado at neighborhoods in close proximity to oil and gas facilities. They found that residents who lived less than half a mile away from wells had a subchronic noncancer health risk that was 25 times greater than for those living further away. This health risk was mainly driven by exposure to trimethylbenzenes, xylenes, and aliphatic (open chain) hydrocarbons. A major flaw in the study of McKenzie et al. (2012) was that no baseline sampling was conducted prior to the onset of local oil and gas development. In contrast to this study, Colborn et al. (2014) collected weekly air samples for one year before, during, and after drilling and hydraulic fracturing at a gas well pad in rural western Colorado. They found numerous hydrocarbons associated with health effects, including 30 endocrine disruptors and some polycyclic aromatic hydrocarbons (PAHs) known to adversely affect prenatal development. The concentrations of these hydrocarbons were highest during the initial drilling phase.

McKenzie et al. (2014) conducted a retrospective cohort study of 124,842 births between 1996 and 2009 in rural Colorado, in which they examined associations between maternal residential proximity to natural gas development and birth outcomes. They found an increased prevalence of congenital heart defects, and possibly neural tube defects, with increasing density of gas wells and increasing proximity of these wells to maternal residences.

REACTIVE CHEMICALS AND MIXTURES

Health effects may not be caused exclusively by primary (i.e., directly emitted) air pollution. Once released to air, pollutants undergo a number of transformations depending on their chemical and physical properties. For example, a compound may undergo photolysis in the presence of sunlight, or if semi-volatile, may partition between the gas phase and particulate matter. Chemical reactions between species may occur involving gaseous, liquid, or even solid phases. Within the gas phase, the most important initial chemical reaction is usually with hydroxyl radical (OH). Radicals are molecular fragments with unpaired electrons that react rapidly with VOCs. Hydroxyl radicals are ubiquitous in the atmosphere and function as cleansers of pollution. Along with photolysis, reactions with OH generally determine the photochemical lifetime of a pollutant, defined as the inverse of the net photolytic plus chemical rate of destruction of the pollutant in air.

Some HAPs such as benzene are long-lived species that do not rapidly react with either sunlight or atmospheric hydroxyl radicals. Those that do, or are themselves the secondary by-product of atmospheric chemical reactions, pose complications in risk assessment, either because their reaction products may also be toxic, or because their primary and secondary sources cannot be easily distinguished in the attempt to attribute exposures to emission activities. For example, formaldehyde is a primary emission from incomplete combustion but is also the secondary by-product of methane and VOC degradation in air. The evaluation of risk due to atmospheric pollution is confounded even further by complex mixtures of chemicals whose interactions in biological receptors are not well understood.

Among the novel exposure assessment techniques that were demonstrated during the BEE-TEX study was the use of in vitro cultured human lung cells to gauge the impact of typical mixtures of chemicals in air, either through the measurement of enzymes and proteins released by the cells when stressed or dying, or the analysis of messenger RNA (mRNA) associated with immune system or cancer response to air pollution (Vizuete et al., 2015). In laboratory studies of cultured lung cell response prior to BEE-TEX, secondary pollutants generated by photochemistry caused a 22-fold increase in the expression of interleukin 8 (IL-8) mRNA relative to control (Doyle et al., 2004, 2007). They also altered the expression of genes that play a role in inflammation response or in cell cycle regulation (Rager et al., 2011). During BEE-TEX, exposure of human lung cells to ambient air pollution near a refinery in the Manchester neighborhood of Houston resulted in differential expression compared to clean air in 10 immune-related genes and 1 cancer-related gene (Vizuete et al., 2015).

On the whole, there remains very limited evidence of direct causal links between oil and gas development and adverse environmental health outcomes, largely because insufficient time has been available for more rigorous studies given the rapid development of oil and gas resources since the turn of the century. Health effects with especially long latencies may not yet be seen. Moreover, exposure science is still in its infancy, and will require more vigorous development and updating with new methods to overcome the challenges posed by the massive increase in the societal use of industrial chemicals. This includes a more complete elucidation of the concept of the "exposome," which refers to the totality of environmental exposures experienced by an individual since conception, as a means to explain disease in a manner that is complementary to the human genome (Egeghy et al., 2015).

REFERENCES

Bunch, A.G., Perry, C.S., Abraham, L., Wikoff, D.S., Tachovsky, J.A., Hixon, J.G., et al., 2014. Evaluation of impact of shale gas operations in the Barnett Shale region on volatile organic compounds in air and potential human health risks. Sci. Total Environ. 468−469, 832−842.

Colborn, T., Schultz, K., Herrick, L., Kwiatkowski, C., 2014. An exploratory study of air quality near natural gas operations. Human Ecol. Risk Assess. 20, 86−105.

Conley, S., Franco, G., Faloona, I., Blake, D.R., Peischl, J., Ryerson, T.B., 2016. Methane emissions from the 2015 Aliso Canyon blowout in Los Angeles, CA. Science. Available from: <http://dx.doi.org/10.1126/science.aaf2348>.

Doyle, M., Sexton, K.G., Jeffries, H.J., Bridge, K., Jaspers, I., 2004. Effects of 1,3-butadiene, isoprene, and their photochemical degradation products on human lung cells. Environ. Health Perspect. 112, 1488−1495.

Doyle, M., Sexton, K.G., Jeffries, H., Jaspers, I., 2007. Atmospheric photochemical transformations enhance 1,3-butadiene-induced inflammatory responses in human epithelial cells: the role of ozone and other photochemical degradation products. Chem. Biol. Interact. 166, 163−169.

Egeghy, P.P., Sheldon, L.S., Isaacs, K.K., Özkaynak, H., Goldsmith, M.-R., Wambaugh, J.F., et al., 2015. Computational exposure science: An emerging discipline to support 21st-century risk assessment. Environ. Health Perspect. Available from: <http://dx.doi.org/10.1289/ehp.1509748>.

Environmental Protection Agency (EPA), 2003. Benzene; CASRN 71-43-2. Integrated Risk Information System (IRIS) chemical assessment summary. Available at: <https://cfpub.epa. gov/ncea/iris/iris_documents/documents/subst/0276_summary.pdf>. Accessed 7/20/2016.

Maddaus, G., 2015. What went wrong at porter ranch? LA Weekly, December 22.

McKenzie, L.M., Guo, R., Witter, R.Z., Savitz, D.A., Newman, L.S., Adgate, J.L., 2014. Birth outcomes and maternal residential proximity to natural gas development in rural Colorado. Environ. Health Perspect. 122, 412−417.

McKenzie, L.M., Witter, R.Z., Newman, L.S., Adgate, J.L., 2012. Human health risk assessment of air emissions from development of unconventional natural gas resources. Sci. Total Environ. 424, 79−87.

Olaguer, E.P., 2015. Overview of the Benzene and other Toxics Exposure (BEE-TEX) field study. Environ. Health Insights 9 (S4), 1−6. Available from: <http://dx.doi.org/10.4137/ EHI.S15654>.

Olaguer, E.P., Erickson, M.H., Wijesinghe, A., Neish, B.S., 2016. Source attribution and quantification of benzene event emissions in a Houston Ship Channel community based on real time mobile monitoring of ambient air. J. Air Waste Manage. Assoc. 66, 164−172.

Rager, J.E., Lichtveld, K., Ebersviller, S., Smeester, L., Jaspers, I., Sexton, K.G., et al., 2011. A toxicogenomic comparison of primary and photochemically altered air pollutant mixtures. Environ. Health Perspect. 119, 1583−1589.

Rich, N., 2016. The Invisible Catastrophe. New York Times Magazine, March 31.

Vizuete, W., Sexton, K.G., Nguyen, H., Smeester, L., Aagaard, K.M., Shope, C., et al., 2015. From the field to the laboratory: Air pollutant-induced genomic effects in lung cells. Environ. Health Insights 9 (S4), 15−23. Available from: <http://dx.doi.org/10.4137/EHI.S15656>.

Werner, A.K., Vink, S., Watt, K., Jagals, P., 2015. Environmental health impacts of unconventional natural gas development: a review of the current strength of evidence. Sci. Total Environ. 505, 1127−1141.

World Health Organization (WHO), 2000. Benzene, Air Quality Guidelines for Europe, 2nd ed World Health Organization Regional Office for Europe, Copenhagen. Available from: <http://www.euro.who.int/__data/assets/pdf_file/0005/74732/E71922.pdf>.

Chapter 4

Urban and Regional Ozone

Chapter Outline

ATMOSPHERIC OZONE AND THE US NATIONAL AMBIENT AIR QUALITY STANDARD

Ozone (O_3) is distributed throughout the earth's atmosphere, which is classified into vertical regions according to variations in temperature and ozone. The troposphere is the region in which the temperature decreases from the earth's surface to a minimum at the tropopause, which at mid-latitudes is around $10-12$ km above ground level (AGL). The stratosphere is the region above the tropopause in which the temperature generally increases with height. Ozone is produced naturally in the stratosphere via the photolysis of oxygen, resulting in stratospheric volume mixing ratios of $1000-10,000$ ppb. Tropospheric ozone concentrations, on the other hand, are generally lower than in the stratosphere by two to three orders of magnitude.

Tropospheric O_3 is not directly emitted by anthropogenic or natural sources, but is a secondary pollutant formed by photochemical reactions of precursor gases, mainly volatile organic compounds (VOCs), and nitrogen oxides (NO_x). Carbon monoxide (CO) also plays an important role in ozone formation in both polluted and remote areas, whereas methane contributes to the formation of global background ozone. The immediate cause of secondary ozone in the troposphere is the photolysis of nitrogen dioxide (NO_2) in the presence of molecular oxygen, but this is only the last step in a complex chain of precursor reactions and chemical by-products. The distribution of ozone in the troposphere is also partly controlled by atmospheric dynamics, including horizontal transport, vertical mixing within the troposphere, and mass exchange between the troposphere and stratosphere.

The largest sources of natural ozone in the troposphere are the mixing of air from the stratosphere, and photochemical reactions involving natural NO_x

Atmospheric Impacts of the Oil and Gas Industry. DOI: http://dx.doi.org/10.1016/B978-0-12-801883-5.00004-8

(from lightning and soils), natural methane (from permafrost, wetlands, forest fires, plants, termites, and oceans), and natural VOCs (from soil microbes and vegetation). These natural processes typically result in 10–25 ppb of O_3 at the surface (Fiore et al., 2003). Occasionally, "tropopause folding" involving the rapid subsidence of stratospheric air can substantially enhance this natural contribution during winter and spring, with much smaller contributions during summer and fall.

The stratospheric ozone layer is vital in screening the surface from dangerous solar ultraviolet (UV) radiation, hence it requires protection from ozone-depleting compounds such as chloro-fluorocarbons (CFCs) and other man-made halogens. Tropospheric ozone enhancement by human activity, on the other hand, is discouraged to protect human health and terrestrial ecology. Breathing ozone at concentrations significantly above natural levels can impair lung function and harm lung tissue, cause symptoms such as chest pain, coughing, throat irritation, and airway inflammation, and aggravate illnesses such as bronchitis, emphysema, and asthma. Elevated tropospheric ozone levels can also damage vegetation, which can lead to losses in species diversity, degradation of habitat, and changes in water and nutrient cycling.

Ozone is one of six air pollutants whose atmospheric concentrations are limited by the US Clean Air Act based on health criteria, hence they are referred to as "criteria pollutants." The other criteria pollutants are CO, NO_2, sulfur dioxide (SO_2), lead (Pb) and particulate matter (PM). These criteria pollutants are each assigned a National Ambient Air Quality Standard (NAAQS), which may include both a primary and a secondary standard. When the NAAQS is violated according to metrics established by the US EPA, a geographical area may be designated in nonattainment of the relevant standard. The legally responsible entity (typically the State) is then required to develop an implementation plan to enable the designated area to eventually reach attainment.

The current primary and secondary standards for ozone are both set at 70 ppb averaged over 8 h. An ozone design value (DV) for each regulatory monitor in a designated area is used to measure nonattainment, and is equal to the annual fourth-highest daily maximum 8-h average concentration, averaged over 3 years. A DV exceeding the NAAQS at any monitor triggers a nonattainment designation and the development of a State Implementation Plan (SIP).

OZONE PHOTOCHEMISTRY

Photochemistry leading to high ozone episodes is triggered by local emissions of NO_x, mostly in the form of NO from combustion sources, and both natural and anthropogenic emissions of VOCs. Although NO at first suppresses ozone in the immediate vicinity of sources due to NO_x titration ($NO + O_3 \rightarrow NO_2 + O_2$), its conversion to NO_2 eventually leads to ozone

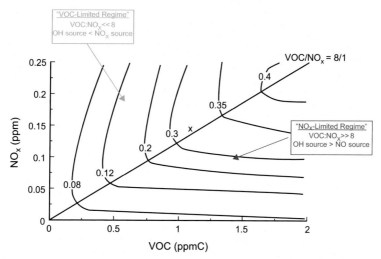

FIGURE 4.1 Ozone isopleths (ppm) illustrating relationship between NO_x, VOCs, and O_3. *Environment Canada and U.S. Environmental Protection Agency, 2014. Georgia Basin—Puget Sound Airshed Characterization Report, 2014. In: Vingarzan, R., So, R., Kotchenruther, R. (Eds.) Environment Canada, Pacific and Yukon Region, Vancouver (BC). U.S. Environmental Protection Agency, Region 10, Seattle, WA. ISBN 978-1-100-22695-8. Cat. No.: En84-3/2013E-PDF. EPA 910-R-14-002.; Dodge, M.C., 1977. Combined use of modeling techniques and smog chamber data to derive ozone -precursor relationships. In: Dimitriades, B. (Ed.), Proceedings of the International Conference of Photochemical Oxidant Pollution and Its Control. EPA-600/3-77001b, Vol II, pp. 881—889.*

production further downwind. The efficiency of ozone production is partly determined by the relative abundances of NO_x and VOCs (see Fig. 4.1). At low VOC to NO_x ratios (< about 4−1), an area is considered to be VOC-limited; VOC reductions will be most effective in reducing ozone, and NO_x controls may lead to ozone increases due to decreased titration near NO_x sources. At high VOC to NO_x ratios (> about 15−1), an area is considered NO_x limited, and VOC controls may be ineffective. There is, however, another factor beyond the VOC to NO_x ratio that determines ozone productivity, specifically the availability of so-called radical precursors.

For ozone to build up to unhealthy levels, there must be a continuous re-cycling between NO and NO_2 fueled by atmospheric radicals, which undergo a parallel cycling in which radicals from primary sources are recovered when NO is converted to NO_2 (see Fig. 4.2). Sources of primary radicals include the photolysis of ozone, aldehydes (RCHO), and nitrous acid (HONO), while the decomposition of VOCs due to reactions with OH provides a source of secondary radicals (RO_x), which then lengthen the radical and NO_x reaction chains that generate ozone. The longer the lengths of these two reaction chains, the more ozone is produced. The radical chain is terminated by reactions of radicals with NO_x or other radicals, while the

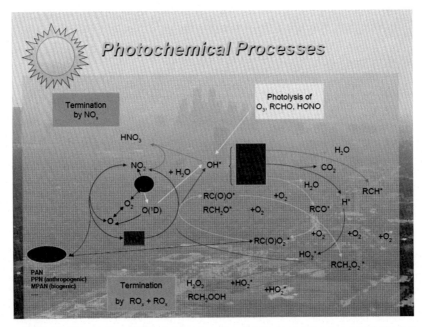

FIGURE 4.2 The cycling of NO_x (left-hand cycle) and radicals (right-hand cycle) that leads to ozone formation. *Olaguer, E.P., Kolb, C.E., Lefer, B., Rappenglück, B., Zhang, R., Pinto, J.P., 2014. Overview of the SHARP campaign: motivation, design, and major outcomes. J. Geophys. Res. Atmos. 119, 2597–2610.*

NO_x chain is terminated by the formation of reactive nitrogen reservoirs such as nitric acid (HNO_3) and peroxyacetyl nitrate (PAN). Without an adequate supply of primary radical precursors that can ignite the radical and NO_x reaction chains, the ozone derived from local emissions of NO_x and VOCs may be significantly limited (Olaguer et al., 2009, 2014).

FACTORS AFFECTING REGIONAL OZONE

Meteorological factors greatly influence ozone pollution in addition to precursor emissions. In the continental United States, high ozone concentrations at ground level during warmer seasons are often associated with slow moving regional high pressure systems (anticyclones). The sinking air within these systems inhibits vertical mixing, promotes higher temperatures and suppresses clouds, thereby enhancing local photochemical activity. Meteorological conditions, however, may also lead to long-range transport of pollution. In practice, this means that local controls may not be adequate to attain the ozone NAAQS, so that broader regional or even continental scale controls may be required.

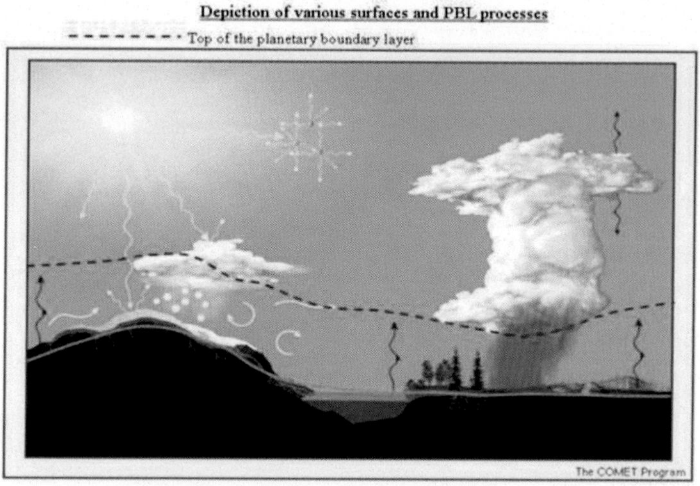

FIGURE 4.3   An illustration of the planetary boundary layer beneath the free troposphere. *National Oceanic and Atmospheric Administration.*

Determining the distribution of local and regional ozone requires an understanding of the dynamics of the planetary boundary layer (PBL), which is the layer of the atmosphere next to the surface within which long-lived pollutants are well mixed (see Fig. 4.3). The PBL height is thus an important determinant of surface ozone concentration, and is typically around 1 km at mid-day or early afternoon, but can grow beyond this on very hot days. At night, the mixing height shrinks to a few hundred meters, trapping air pollution much closer to the ground. Deep convection during warm days produces cumulus towers that rapidly vent pollution from the PBL to the free troposphere above it. Due to mass conservation, deep convection also forces air around the cumulus towers to slowly subside, entraining air pollutants back into the PBL. The importance of convection in maintaining the global tropospheric ozone cycle was demonstrated by Lelieveld and Crutzen (1990).

Because dry deposition is restricted to the PBL, ozone has a longer lifetime in the free troposphere, and can be transported long distances together with other pollutants by large-scale circulation systems, such as the jet stream, mid-latitude cyclones (regional low pressure systems), and "warm conveyor belts" (see Fig. 4.4) associated with frontal activity (Cooper and Parrish, 2004).

Some ozone aloft may have been formed within urban plumes far upwind or from urban NO_x emissions and biogenic VOCs (e.g., isoprene) emitted by regional scale vegetation. Prolonged atmospheric stability and drier air aloft can allow pollutants in the lower free troposphere to persist for multiple days. Rapid photochemical processing within the dirty air pool converts short-lived NO_x into longer-lived reactive nitrogen reservoirs such as PAN.

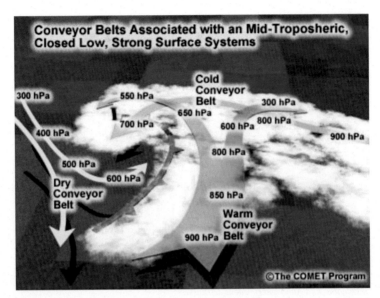

FIGURE 4.4 Conveyor belts associated with a maturing cyclone, with pressure levels denoted in millibars or hectopascals (hPa). *US National Weather Service.*

When this pollution is entrained into the PBL and the nitrogen reservoirs decompose, the NO_x is recovered and can be converted to ozone. Analyses by Taubman et al. (2004, 2006) of airborne measurements in the eastern United States during multi-day pollution episodes demonstrated that high surface ozone may be accompanied by high concentrations of ozone and other pollutants aloft that are traceable to distant sources.

Long-range transport of air pollution is not limited to the daytime. Concentrated ozone plumes aloft are sometimes observed at night far from the original source of precursors due to the suppression of vertical mixing and the presence of nocturnal low level jets a few hundred meters above ground. Banta et al. (2005) concluded that during the 2000 Texas Air Quality Study (TexAQS I), the lofting of large quantities of pollution in the PBL above Houston may have influenced air quality on regional scales through nocturnal transport.

Long-range transport of ozone and its precursors must be taken into account in setting the NAAQS. For this purpose, the US. EPA (2013) defines background ozone in a manner that distinguishes natural, North American and United States background concentrations. This facilitates the separation of pollution levels that can be controlled by domestic regulations (or through international agreements with neighboring countries) from levels that are generally uncontrollable by the United States. In particular, United States background concentrations are defined by simulating the O_3 concentrations that would exist in the absence of anthropogenic emissions from the

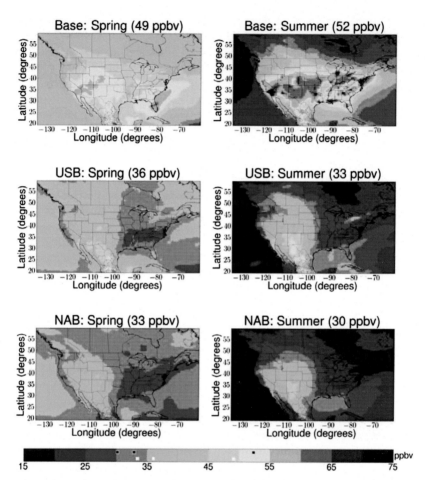

FIGURE 4.5   Seasonal mean daily maximum 8-h average O_3 concentrations calculated by GEOS-Chem for the base case (top, Base), United States background (middle, USB) and North American background (bottom, NAB). Values in parentheses refer to means, and are shown in the color bar as black squares for summer and white squares for spring. *Environmental Protection Agency (EPA), 2013. Integrated science assessment for ozone and related photochemical oxidants. EPA 600/R-10/076F. Office of Research and Development, Research Triangle Park, NC.*

continental United States (CONUS), whereas North American background concentrations account only for anthropogenic emissions outside the CONUS, Canada, and Mexico. Based on simulations with the GEOS-Chem global chemical transport model (Zhang et al., 2011), average United Ststes background concentrations are 36 ppb by volume in spring and 33 ppb in summer, while corresponding North American background concentrations are about 3 ppb lower (see Fig. 4.5).

For ozone attainment demonstrations, it is useful to define background ozone on a smaller scale as compared to the EPA definition. Nielsen-Gammon et al. (2005) defined background ozone for an urban area as the concentration level that would exist if there were no local anthropogenic precursor emissions. Operationally, they estimated background ozone as the lowest maximum 8-h ozone concentration observed on the outskirts (typically upwind) of the urban area. Applying this method to all days over a six-year period for various cities of eastern Texas, Nielsen-Gammon et al. (2005) showed that background ozone in eastern Texas peaks during the spring and late summer, and ranges from 20 to 55 ppb.

During the 2000 TexAQS I and 2006 Second Texas Air Quality Study (TexAQS II), aircraft measured ozone concentrations in the vicinity of the Houston and Dallas–Fort Worth (DFW) metropolitan areas. Analyses of these aircraft measurements by Kemball-Cook et al. (2009) showed that these two urban regions could be brought close to exceeding the then-prevailing 1997 8-h NAAQS for ozone of 80 ppb solely due to regional transport, without considering the contribution of local sources.

OIL AND GAS DEVELOPMENT AND URBAN OZONE NONATTAINMENT

In making the current 2015 ozone NAAQS more stringent than its 1997 and 2008 predecessors, the EPA reasoned that lowering the standard would generate nationwide health benefits in 2025 (outside of California, which would take longer to comply with the NAAQS) of $2.9–5.9 billion, with costs of $1.4 billion (McCarthy and Lattanzio, 2016). In addition, it estimated that crop losses could be reduced by $400–620 million. The revised ozone standard leaves very little room for ozone produced by local emission sources during episodes of high regional background ozone. This fact may have severe consequences for oil and gas development in nonattainment areas such as DFW, which sits atop the Barnett Shale.

The ozone problem in DFW is different from that of more notorious airsheds, such as Los Angeles and Houston. Los Angeles is situated between a mountain range and the Pacific Ocean, so that local pollution can be trapped by the incoming sea breeze and warm subsiding air that creates a temperature ("capping") inversion aloft. Houston is situated near the Gulf of Mexico. The Gulf sea breeze cycle creates a rotating wind pattern that allows pollution generated by numerous petrochemical facilities along the Houston Ship Channel, as well as traffic emissions, to stagnate in place (Olaguer et al., 2006). DFW, however, is located on flat terrain far inland.

The conceptual model for DFW ozone exceedances developed by the Texas Commission on Environmental Quality (TCEQ, 2015) acknowledges that high local ozone concentrations tend to occur when regional background ozone is also elevated, especially with slow winds from the east or southeast.

TABLE 4.1 2009 Peak Summer Emissions From Barnett Shale Oil and Gas Activities

Source	Emissions (tons per day)	
	NO$_x$	VOC
Compressor engine exhaust	46	19
Condensate and oil tanks	0	146
Production fugitives	0	26
Well drilling and completions	5.5	21
Gas processing	0	15
Transmission fugitives	0	28
Total emissions	51	255

Source: Armendariz, A., 2012. Emissions from natural gas production in the Barnett Shale area and opportunities for cost-effective improvements. Report to the Environmental Defense Fund, Austin, TX.

This makes DFW susceptible to influences from elsewhere in Texas and beyond, including emissions from large power plants or urban plumes from major cities such as Houston (Olaguer et al., 2006). But it also raises the question as to whether much nearer oil and gas activity in the Barnett Shale may aggravate ozone nonattainment.

Armendariz (2012) estimated ozone precursor emissions from oil and gas production in the Barnett Shale for 2009 and compared these to emissions from other source categories in the DFW area (see Table 4.1 and Fig. 4.6). His results showed that emissions from oil and gas activities could exceed mobile source emissions in the same region. The TCEQ has since compiled a special inventory for stationary oil and gas sources, but did not include emissions from well drilling and completion activities (Ethridge et al., 2015).

Ahmadi and John (2015) conducted statistical analyses of 14 years of hourly ozone and meteorological data collected at regulatory monitoring sites in the Barnett Shale. In doing so, they divided the DFW area into two regions, the shale gas region (SGR) and the nonshale gas region (NSGR), according to the number of gas wells in close proximity to monitoring sites. They found that monitoring sites in the NSGR showed a greater reduction in ozone than those in the SGR, and that winds blowing from areas with high shale gas activities contributed to higher ozone downwind.

Although the study of Ahmadi and John (2015) has yielded some tantalizing indications, the density of ozone monitors in DFW may still not be

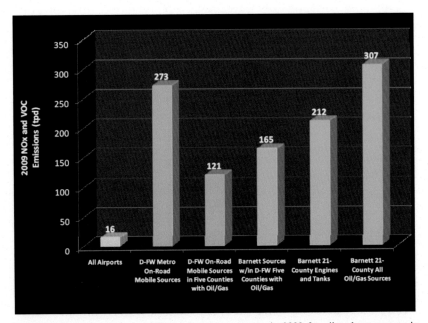

FIGURE 4.6 Peak summer emissions of ozone precursors in 2009 for oil and gas sources in the Barnett Shale and other source categories in the DFW region. *Armendariz, A., 2012. Emissions from natural gas production in the Barnett Shale area and opportunities for cost-effective improvements. Report to the Environmental Defense Fund, Austin, TX.*

sufficient to detect many influences from oil and gas activities in the Barnett Shale. As previously discussed in Chapter 1, A Brief History of Oil and Gas Development From an Environmental Perspective, the micro-scale modeling study of Olaguer (2012) showed that compressor engines and flares burning natural gas may generate a few ppb of ozone only a few kilometers downwind of oil and gas sites.

In the case of natural gas flares, the resulting ozone impacts may be heightened by incomplete combustion that generates direct emissions of highly reactive olefins (hydrocarbons with at least one double bond, as compared to alkanes which have only single bonds) and the radical precursor, formaldehyde. These emissions are currently ignored in official inventories, which typically assume that flares operate at 99% combustion efficiency, and that the VOCs emitted amount to 1% of the corresponding hydrocarbon (primarily alkane) flows to the flare independent of flare combustion chemistry. This oversimplification is partly due to the lack of monitoring studies focused on upstream natural gas flaring, as compared to downstream petrochemical flares. The TCEQ conducted its own experimental study of flare emissions in 2010, which showed that air and steam-assisted flares that burn a mixture of 80% propane and 20% Tulsa Natural Gas can produce significant amounts of formaldehyde and olefins (Knighton et al., 2012). The use of air and steam assist significantly lowers the combustion efficiency of flares, leading to greater emissions of products of incomplete combustion. However, even flares that are not steam or air-assisted may have their combustion efficiencies reduced below 90% by moderate crosswinds (Castineira, 2006; Castineira and Edgar, 2008).

Combined emissions of formaldehyde and olefins from flares and other industrial sources have been demonstrated to play a significant role in ozone formation in the Houston airshed, based on the results of the TexAQS II field study in 2006 (Olaguer et al., 2009) and the Study of Houston Atmospheric Radical Precursors (SHARP) in 2009 (Olaguer et al., 2014). While emissions of these ozone precursors from the upstream oil and gas industry may not be as large as for the downstream sector, their corresponding air quality impacts may nevertheless be significant.

Schade and Roest (2016) analyzed a year's worth of hydrocarbon and other measurements (not including formaldehyde) from a Continuous Ambient Monitoring Station (CAMS) operated by the TCEQ in Floresville, Texas, at the central north edge of the Eagle Ford Shale. They identified two dominant factors in the ambient air data. The first factor had a relative contribution between 47% and 50% and was assigned to emissions from oil and gas activities due to the dominance of alkanes in the factor composition. The second factor had a relative contribution between 29% and 32% and was dominated by olefins and acetylene, with large contributions from NO_x and aromatic hydrocarbons (i.e., compounds containing one or more benzene rings). This second factor was assigned to combustion sources, which may include urban traffic emissions as well as flaring and engine emissions from oil and gas sites. An analysis of factor variability with wind direction by Schade and Roest (2016) provided evidence that natural gas combustion associated with shale petroleum mining may have contributed significantly to the observed loadings of light olefins such as ethene and propene, which were sometimes correlated with alkanes.

THE COLD OZONE PHENOMENON

Severe smog is usually produced under hot and humid urban conditions. However, much to the surprise of many, cold conditions do not prevent such smog from appearing in rural areas where oil and gas development is prevalent. Schnell et al. (2009) first reported rapid formation of ozone with air temperatures as low as $-17°C$ at the Jonah–Pinedale Anticline natural gas field in Wyoming's Upper Green River Basin. They found that hourly average ozone during the winter of 2008 rose from 10 to 30 ppb at night to more than 140 ppb in the early afternoon. A stagnant, high-pressure system, low

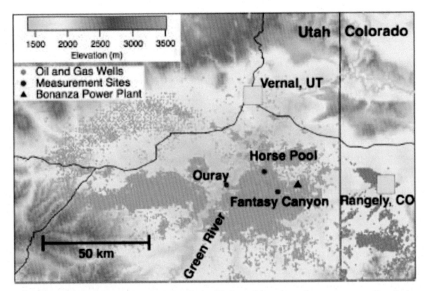

FIGURE 4.7 Topography of the Uinta Basin, Utah, and location of measurement sites during the UBWOS field campaigns. *Ahmadov, R., McKeen, S., Trainer, M., Banta, R., Brewer, A., Brown, S., et al., 2015. Understanding high wintertime ozone pollution events in an oil- and natural gas-producing region of the western US. Atmos. Chem. Phys. 15, 411−429.*

wind speeds, and relatively clear skies enabled an intense temperature inversion to develop within a 100 m thick layer above the surface, which trapped high concentrations of ozone precursors. Daytime photolysis amplified by strong reflection of sunlight by snow cover then led to the rapid rise in ozone mixing ratio. This phenomenon was repeated in the winter of 2011, and as of July 2012, the EPA designated the Upper Green River Basin as an ozone nonattainment area (Rappenglück et al., 2014).

The cold ozone phenomenon has likewise been observed in the Uinta Basin of Utah, where rapid oil and gas development has also taken place. This was the subject of the Uinta Basin Winter Ozone Studies (UBWOS) conducted by the National Oceanic and Atmospheric Administration (NOAA) and other researchers in 2012, 2013, and 2014 (see Fig. 4.7).

Edwards et al. (2014) analyzed UBWOS measurements for February 2013 using a simple box model with a detailed photochemical mechanism and emissions tuned to match observed ambient concentrations. They successfully reproduced the temporal behavior of ozone and various precursors (see Fig. 4.8), and concluded that winter ozone production in the Uinta Basin occurs at lower NO_x and much larger VOC concentrations than in urban areas during summer. Edwards et al. (2014) also found that photolysis of carbonyls (compounds with a carbon−oxygen double bond) was a dominant

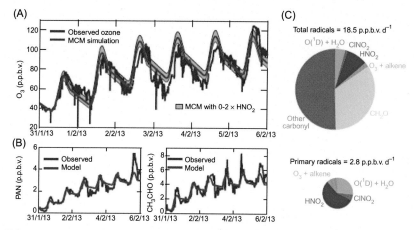

FIGURE 4.8 Box model simulations of (A) ozone, (B) PAN and acetaldehyde with the Master Chemical Mechanism (MCM) versus corresponding observations at the Horse Pool site during UBWOS 2013. Also shown in (C) are various contributions to the primary and total radical budget. *Edwards, P.M., Brown, S.S., Roberts, J.M., Ahmadov, R., Banta, R.M., deGouw, J.A., et al., 2014. High winter ozone pollution from carbonyl photolysis in an oil and gas basin. Nature 514, 351—354.*

source of radicals in the Uinta Basin, unlike in urban areas where ozone photolysis in the presence of water vapor is the largest contributor to the radical budget during high ozone episodes. However, they under predicted formaldehyde (one of the simplest carbonyl species) by 30%, since their model did not include direct emissions of this compound from oil and gas sites, so that formaldehyde (CH_2O in Fig. 4.8) was considered a secondary (rather than primary) radical source.

Ahmadov et al. (2015) analyzed UBWOS data using a 3D regional chemical transport model and an explicit high resolution simulation of the meteorology of the Uinta Basin. A model simulation with bottom-up emission estimates derived from the 2011 National Emissions Inventory (NEI) underestimated methane and VOC concentrations, overestimated NO_x concentrations, and failed to reproduce any of the observed high ozone episodes during UBWOS. On the other hand, a top-down emission inventory derived from *in situ* aircraft measurements successfully captured the observed buildup and afternoon peaks in ozone. The results of Ahmadov et al. (2015) showed a disproportionate contribution of aromatic hydrocarbons to ozone formation compared to all other VOCs. However, their simulations underestimated ambient formaldehyde concentrations by 50% even with direct emissions assumed by the model. Because of the sensitivity of Uinta Basin ozone to the production of radicals by formaldehyde photolysis, they stressed the importance of quantifying primary versus secondary formaldehyde in further studies.

REFERENCES

Ahmadi, M., John, K., 2015. Statistical evaluation of the impact of shale gas activities on ozone pollution in North Texas. Sci. Total Environ. 536, 457–467.

Ahmadov, R., McKeen, S., Trainer, M., Banta, R., Brewer, A., Brown, S., et al., 2015. Understanding high wintertime ozone pollution events in an oil- and natural gas-producing region of the western US. Atmos. Chem. Phys. 15, 411–429.

Armendariz, A., 2012. Emissions from natural gas production in the Barnett Shale area and opportunities for cost-effective improvements. Report to the Environmental Defense Fund, Austin, TX.

Banta, R.M., Senff, C.J., Nielsen-Gammon, J., Darby, L.S., Ryerson, T.B., Alvarez, R.J., et al., 2005. A bad air day in Houston. B. Am. Meteorol. Soc. 86, 657–669.

Castineira, D., 2006. A Computational Fluid Dynamics Simulation Model for Flare Analysis and Control (Ph.D. thesis). University of Texas at Austin.

Castineira, D., Edgar, T.F., 2008. C.F.D. for simulation of crosswind on the efficiency of high momentum jet turbulent combustion flames. J. Environ. Eng. 134, 561–571.

Cooper, O.R., Parrish, D.D., 2004. Air pollution export from and import to North America: experimental evidence. Handb. Environ. Chem. 4, Part G41–67.

Dodge, M.C., 1977. Combined use of modeling techniques and smog chamber data to derive ozone -precursor relationships. In: Dimitriades, B. (Ed.), Proceedings of the International Conference of Photochemical Oxidant Pollution and Its Control, Vol II. US EPA, Office of Research and Development, Research Triangle Park, NC, pp. 881–889. EPA-600/3-77001b.

Edwards, P.M., Brown, S.S., Roberts, J.M., Ahmadov, R., Banta, R.M., deGouw, J.A., et al., 2014. High winter ozone pollution from carbonyl photolysis in an oil and gas basin. Nature 514, 351–354.

Environmental Protection Agency (EPA), 2013. Integrated Science Assessment for Ozone and Related Photochemical Oxidants. EPA 600/R-10/076F. Office of Research and Development, Research Triangle Park, NC.

Environment Canada and U.S. Environmental Protection Agency, 2014. Georgia Basin—Puget Sound Airshed Characterization Report, 2014. In: Vingarzan, R., So, R., Kotchenruther, R. (Eds.), Environment Canada, Pacific and Yukon Region, Vancouver (BC). U.S. Environmental Protection Agency, Region 10, Seattle, WA, ISBN 978-1-100-22695-8. Cat. No.: En84-3/2013E-PDF. EPA 910-R-14-002.

Ethridge, S., Bredfeldt, T., Sheedy, K., Shirley, S., Lopez, G., Honeycutt, M., 2015. The Barnett Shale: from problem formulation to risk management. J. Unconventional Oil Gas Resour. 11, 95–110.

Fiore, A., Jacob, D.J., Liu, H., Yantosca, R.M., Fairlie, T.D., Li, Q., 2003. Variability in surface ozone background over the United States: implications for air quality policy. J. Geophys. Res. 108 (D24), 4787. Available from: <http://dx.doi.org/10.1029/2003JD003855>.

Kemball-Cook, S., Parrish, D., Ryerson, T., Nopmongcol, U., Johnson, J., Tai, E., et al., 2009. Contributions of regional transport and local sources to ozone exceedances in Houston and Dallas: comparison of results from a photochemical grid model to aircraft and surface measurements. J. Geophys. Res. 114, D00F02. Available from: <http://dx.doi.org/10.1029/2008JD010248>.

Knighton, W.B., Herndon, S.C., Franklin, J.F., Wood, E.C., Wormhoudt, J., Brooks, W., et al., 2012. Direct measurement of volatile organic compound emissions from industrial flares using real-time online techniques: Proton Transfer Reaction Mass Spectrometry and Tunable Infrared Laser Differential Absorption Spectroscopy. Ind. Eng. Chem. Res. 51, 12674–12684.

Lelieveld, J., Crutzen, P.J., 1990. Role of deep cloud convection in the ozone budget of the troposphere. Science 264, 1759–1761.

McCarthy, J.E., Lattanzio, R.K., 2016. Ozone Air Quality Standards: EPA's 2015 Revision. Congressional Research Service, Washington, DC.

Nielsen-Gammon, J., Tobin, J., McNeel, A., 2005. A conceptual model for eight-hour ozone exceedances in Houston, Texas. Part I: background ozone levels in eastern Texas. Project H12.8HRA Report. Texas Environmental Research Consortium, Houston, TX.

Olaguer, E.P., 2012. The potential near source ozone impacts of upstream oil and gas industry emissions. J. Air Waste Manag. Assoc. 62, 966–977.

Olaguer, E.P., Jeffries, H., Yarwood, G., Pinto, J., 2006. Attaining the 8-hour standard in East Texas: a tale of two cities. EM 8, 26–30.

Olaguer, E.P., Rappenglück, B., Lefer, B., Stutz, J., Dibb, J., Griffin, R., et al., 2009. Deciphering the role of radical precursors during the Second Texas Air Quality Study. J. Air Waste Manag. Assoc. 59, 1258–1277.

Olaguer, E.P., Kolb, C.E., Lefer, B., Rappenglück, B., Zhang, R., Pinto, J.P., 2014. Overview of the SHARP campaign: motivation, design, and major outcomes. J. Geophys. Res. Atmos. 119, 2597–2610.

Rappenglück, B., Ackermann, L., Alvarez, S., Golovko, J., Buhr, M., Field, R.A., et al., 2014. Strong wintertime ozone events in the Upper Green River basin, Wyoming. Atmos. Chem. Phys. 14, 4909–4934.

Schade, G.W., Roest, G., 2016. Analysis of non-methane hydrocarbon data from a monitoring station affected by oil and gas development in the Eagle Ford shale, Texas. Elementa: Sci. Anthrop. 4. Available from: <http://dx.doi.org/10.12952/journal.elementa.000096>.

Schnell, R.C., Oltmans, S.J., Neely, R.R., Endres, M.S., Molenar, J.V., White, A.B., 2009. Rapid photochemical production of ozone at high concentrations in a rural site during winter. Nat. Geosci. 2, 120–122.

Taubman, B.F., Marufu, L.T., Piety, C.A., Doddridge, B.G., Stehr, J.W., Dickerson, R.R., 2004. Airborne characterization of the chemical, optical, and meteorological properties and origins of a combined ozone-haze episode over the eastern United States. J. Atmos. Sci. 61, 1781–1793.

Taubman, B.F., Hains, J.C., Thompson, A.M., Marufu, L.T., Doddridge, B.G., Stehr, J.W., et al., 2006. Aircraft vertical profiles of trace gas and aerosol pollution over the mid-Atlantic U.S.: statistics and meteorological cluster analysis. J. Geophys. Res. 111 (D10), D10S07. Available from: <http://dx.doi.org/10.1029/2005JD006196>.

Texas Commission on Environmental Quality (TCEQ), 2015. Appendix D: conceptual model for the DFW attainment demonstration SIP revision for the 1997 eight-hour ozone standard. In: Revisions to the State of Texas Air Quality Implementation Plan for the Control of Ozone Air Pollution, Austin, TX, June 3.

Zhang, L., Jacob, D.J., Downey, N.V., Wood, D.A., Blewitt, D., Carouge, C.C., et al., 2011. Improved estimate of the policy-relevant background ozone in the United States using the GEOS-Chem global model with 1/2 2/3 horizontal resolution over North America. Atmos. Environ. 45, 6769–6776.

Chapter 5

Particulate Matter and Surface Deposition

Chapter Outline

PARTICULATE MATTER STANDARDS

Particulate matter (PM) pollution consists of solid particles and liquid droplets in air, and may include mixtures of organics, acids, metals, minerals, and elemental carbon. Larger particles such as soil, sand or dust may be seen by the naked eye, while others are microscopic in size. PM that is regulated is at least seven times smaller than the width of human hair. Particles with diameters less than 10 micrometers (μm or microns) are inhalable and may have serious health effects; they are referred to collectively as PM_{10} pollution. Particles with diameters less than 2.5 μm are referred to as $PM_{2.5}$ pollution or fine particulates, in contrast to coarse particulates which have diameters greater than 2.5 μm. Fig. 5.1 illustrates the relative distribution of PM according to size, and the physical transformations that contribute to particle growth.

Health effects of PM include respiratory symptoms (airway irritation, coughing, aggravated asthma, and decreased lung function), cardiovascular problems (heart attacks, irregular heartbeat), cancer, premature delivery, birth defects, and premature death. Smaller particles pose a greater danger of penetrating other organs beyond the lungs. For example, soot nanoparticles (diameter less than 0.1 μm) emitted by diesel engines may be coated with carcinogens such as Polycyclic Aromatic Hydrocarbons (PAHs) and enter into the brain (Donaldson et al., 2005; Prajapati and Tripathi, 2008).

PM has other adverse impacts. Fine particles are the main cause of haze, which reduces visibility in urban areas. Due to the influence of long range transport, haze can also affect otherwise pristine areas such as national parks and wilderness. When deposited to the surface, PM can change the acidity and nutrient balance in soil and surface waters, damage vegetation and affect

Atmospheric Impacts of the Oil and Gas Industry. DOI: http://dx.doi.org/10.1016/B978-0-12-801883-5.00005-X
47

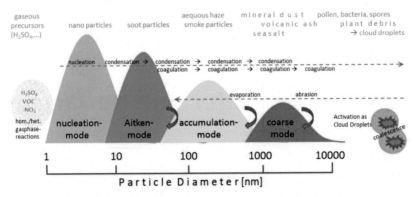

FIGURE 5.1   Schematic illustration of particle types, transformations, and size distributions associated with various modes. *Deutscher Wetterdienst.*

TABLE 5.1 2012 National Ambient Air Quality Standards for PM

Size	Type	Averaging Time	Standard ($\mu g/m^3$)	Form
$PM_{2.5}$	Primary	Annual	12.0	Annual arithmetic mean, averaged over 3 years
	Secondary	Annual	15.0	Annual arithmetic mean, averaged over 3 years
	Primary and secondary	24-hour	35	98th percentile, averaged over 3 years
PM_{10}	Primary and secondary	24-hour	150	Fourth highest 24-h arithmetic mean averaged over 3 years

Source: USEPA.

the diversity of ecosystems. Particle pollution can also cause aesthetic damage to buildings and culturally important objects, and contribute to climate change by altering the radiation balance of the earth's atmosphere.

To protect human health and welfare, the USEPA has established National Ambient Air Quality Standards (NAAQS) for both $PM_{2.5}$ and PM_{10}. The current PM standards are listed in Table 5.1, along with the relevant design value metrics.

PRIMARY PARTICULATES FROM OIL AND GAS ACTIVITIES

Oil and gas activities emit particulates as well as gaseous pollutants. Roy et al. (2014) constructed an air quality emission inventory for the Marcellus Shale region, which lies in the states of Ohio, West Virginia, Pennsylvania,

New York, and Maryland. The city of Pittsburgh, Pennsylvania is located within an area atop the Marcellus Shale that is currently designated as a moderate nonattainment area for $PM_{2.5}$ and may be influenced by PM emissions from oil and gas sources. Tables 5.2 and 5.3 summarize $PM_{2.5}$ emissions estimated by Roy et al. (2014) for a baseline year of 2009 and corresponding projections for the year 2020.

Fig. 5.2 compares Marcellus Shale oil and gas development emissions of $PM_{2.5}$ estimated by Roy et al. (2014) to all other sources in the same geographical region. While the figure indicates that Marcellus development contributes negligibly to overall regional emissions of $PM_{2.5}$, the oil and gas industry may be a relatively important source for some $PM_{2.5}$ components. In particular, Roy et al. (2014) concluded that Marcellus development could contribute 14% of regional elemental carbon (soot) emissions based on diesel source profiles from the EPA (2006) SPECIATE database.

TABLE 5.2 Process-level estimates for $PM_{2.5}$ emissions in the Marcellus Shale for the years 2009 and 2020

Source	Emissions for 2009	Emissions for 2020
Drill rigs	0.3 tons/well drilled	0.1 tons/well drilled
Fracking pumps	0.16 tons/well drilled	0.1 tons/well drilled
Trucks	0.07 tons/well drilled	0.02 tons/well drilled
Compressor stations	0.3 tons/billion cubic feet	0.3 tons/billion cubic feet

Source: Roy, A.A., Adams, P.J., Robinson, A.L., 2014. Air pollutant emissions from the development, production, and processing of Marcellus Shale natural gas. J. Air Waste Manage. Assoc. 64, 19–37.

TABLE 5.3 Activity-level estimates for $PM_{2.5}$ emissions in the Marcellus Shale for the years 2009 and 2020

Source	Emissions for 2009	Emissions for 2020
Well development	0.5 tons/well drilled	0.2 tons/well drilled
Production	0.01 tons/producing well	0.01 tons/producing well
Midstream	0.3 tons/billion cubic feet	0.3 tons/billion cubic feet

Source: Roy, A.A., Adams, P.J., Robinson, A.L., 2014. Air pollutant emissions from the development, production, and processing of Marcellus Shale natural gas. J. Air Waste Manage. Assoc. 64, 19–37.

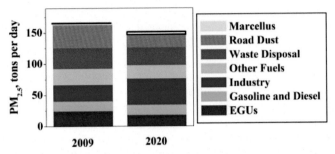

FIGURE 5.2 PM$_{2.5}$ emissions in the Marcellus Shale region from various source categories estimated for the years 2009 and 2020. Oil and gas development emissions are colored yellow. The acronym EGU stands for electricity generating unit. *Roy, A.A., Adams, P.J., Robinson, A.L., 2014. Air pollutant emissions from the development, production, and processing of Marcellus Shale natural gas. J. Air Waste Manage. Assoc. 64, 19–37.*

One particular type of PM pollution stands out with regard to hydraulic fracturing emissions. Esswein et al. (2013) called attention specifically to the hazards of using crystalline silica sand as a proppant to open cracks in shale rock. Workers at oil and gas sites are exposed to these hazards by trucks that transport silica sand to drill sites, and by processes that mix the sand into fracturing fluids. Silica sand may cause silicosis, which involves inflammation and scarring of lung tissue, resulting in diminished lung function. Inhaling silica is also linked to other diseases, such as chronic obstructive pulmonary disease, tuberculosis, kidney disease, autoimmune disorders, and lung cancer (CDC, 2002). Esswein et al. (2013) determined worker exposure to silica sand by collecting air samples at eleven sites in five states. They concluded that the majority of air samples exceeded the permissible exposure level determined by the Occupational Safety and Health Administration (OSHA), as well as the exposure limit recommended by the National Institute for Occupational Safety and Health (NIOSH).

SECONDARY AEROSOL

An important component of PM pollution is secondary aerosol formed from chemical reactions of VOCs, NO$_x$, sulfur compounds, and other atmospheric constituents. Primary PM is generally composed of mineral dust, combustion soot (black or elemental carbon), biogenic aerosol and sea salt. Secondary aerosols are subdivided into inorganic and organic components. Secondary inorganic aerosol results from the well-established chemistry of a few species, resulting in relatively inert ammonium sulfate, ammonium nitrate, and ammonium chloride compounds. Secondary organic aerosol (SOA), on the other hand, involves thousands of compounds that react in multiple generations to form unstable oxidation products, the chemistry of which is only barely understood. Current theory subdivides SOA into hydrocarbon-like organic aerosol (HOA), which has low oxygen content and may have a primary contribution, and oxidized organic aerosol (OOA) with higher oxygen content (Zhang et al., 2005).

Organic aerosols undergo an aging process as they are chemically transformed and grow in the atmosphere. Close to sources, especially in urban areas, a large percentage of these aerosols will consist of smaller HOA. The percentage of HOA will progressively decrease as the SOA is transported downwind, until the HOA component is reduced to a minor fraction in remote areas. The OOA component will not only increase in oxidation state, but also become progressively less volatile and more likely to absorb water vapor (Jimenez et al., 2009; Massoli et al., 2010), eventually becoming subject to wet deposition and removal from the atmosphere.

An important concept in describing SOA evolution is the so-called mixing state of an aerosol population, which Riemer and West (2013) define as "the distribution of the per-particle chemical species compositions." An external mixture consists of particles that each contain only one species (which may be different for each particle), whereas an internal mixture is an aerosol population in which each particle contains multiple species. A fully internal mixture is one in which all particles have the same species composition and relative abundances. As SOA ages, it progresses towards a more internally mixed state. Mixing state is important in that it can affect the physical and optical properties of aerosols (Wang et al., 2010), which determine their effect on the hydrological cycle and climate through the formation of cloud condensation nuclei (CCN) and radiative forcing (see Fig. 5.3).

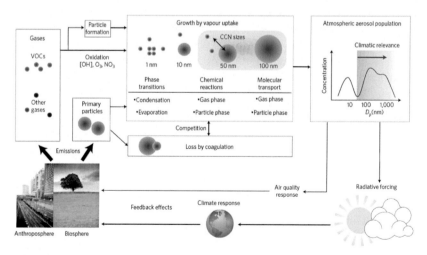

FIGURE 5.3 The cycle of aerosol growth in the atmosphere and its impacts on air quality and climate. *Riipinen, I., Yli-Juuti, T., Pierce, J.R., Tuukka Petaja, T., Worsnop, D.R., Kulmala, M., et al., 2012. The contribution of organics to atmospheric nanoparticle growth. Nat. Geosci. 5, 453–458.*

Lee et al. (2015) investigated the secondary organic nitrate by-products of alkane emissions from oil and gas sites during the 2012 phase of the Uintah Basin Winter Ozone Studies (UBWOS). They found that condensation of gas-phase nitrates and heterogeneous reactions of NO_3 and N_2O_5 resulted in a large fraction of organic nitrates in fine aerosol mass. Lee et al. (2015) concluded that organic nitrate production from alkanes may be a major SOA source.

PARTICLE DEPOSITION AND ECOSYSTEM IMPACTS

Although human health has been the dominant driver for regulating PM pollution, the deposition of particulates onto sensitive ecological receptors may be a significant long-term problem. Atmospheric deposition of PM takes place through three major routes: (1) wet deposition by precipitation scavenging, which is most effective for fine particulates; (2) slower dry deposition, mainly through coarse PM; and (3) occult deposition by fog, cloud water, and mist. Fine particles travel much further than coarse particles before being deposited onto the earth's surface.

Grantz et al. (2003) review what is known about PM effects on ecosystems, and only a brief summary is provided here. Ecological influences include acidic or acidifying nitrate and sulfate (i.e., acid rain), and trace elements and heavy metals. Sulfate and nitrate are mostly present on fine particles, while base cations and heavy metals are largely borne by coarse particles. PM deposition of nitrogen compounds may contribute to excessive algae growth (eutrophication) in surface waters (Paerl, 1997), as well as influence the balance of nutrients in soil. PM may also act as a carrier for phytotoxic materials. Deposition of PM on foliar surfaces may affect vegetation directly, while indirect influences may result from changes in soil chemistry (through altered nutrient cycling and uptake) or in the surface radiation balance. The ecosystem stresses resulting from PM deposition may decrease biodiversity, reduce production, and increase the prevalence of disease.

The oil and gas industry is unlikely to cause large-scale ecological impacts through PM deposition, given that other source categories dominate the budget of PM emissions. However, site-specific and constituent-specific impacts downwind of oil and gas facilities may be possible. For example, Kostiuk et al. (2004) concluded that flaring of propane-rich gas may result in emissions of soot particles to which several PAHs may be attached, including naphthalene, acenapthalene, fluorine, phenanthrene, fluoranthene, and pyrene. Atmospheric deposition is a major source for PAHs in soil, which may be ingested by mammals and readily absorbed from gastrointestinal tracts due to their high lipid solubility (Abdel-Shafy and Mansour, 2016).

REFERENCES

Abdel-Shafy, H.I., Mansour, M.S.M., 2016. A review on polycyclic aromatic hydrocarbons: source, environmental impact, effect on human health and remediation. Egypt. J. Petrol. 25, 107–123.

Centers for Disease Control and Prevention (CDC), 2002. Health effects of occupational exposure to respirable crystalline silica. NIOSH 2002-129. Available at: <http://www.cdc.gov/niosh/docs/2002-129/>.

Donaldson, K., Tran, L., Jimenez, L.A., Duffin, R., Newby, D.E., Mills, N., et al., 2005. Combustion-derived nanoparticles: a review of their toxicology following inhalation exposure. Part. Fibre Toxicol. 2. Available from: http://dx.doi.org/10.1186/1743-8977-2-10.

Environmental Protection Agency (EPA), 2006. SPECIATE 4.0: speciation database development documentation. Available at: <http://www.epa.gov/ttnchie1/software/speciate/>.

Esswein, E.J., Breitenstein, M., Snawder, J., Kiefer, M., Sieber, W.K., 2013. Occupational exposures to respirable crystalline silica during hydraulic fracturing. J. Occup. Environ. Hyg. 10, 347–356.

Grantz, D.A., Garner, J.H.B., Johnson, D.W., 2003. Ecological effects of particulate matter. Environ Int 29, 213–239.

Jimenez, J.L., Canagaratna, M.R., Donahue, N.M., et al., 2009. Evolution of organic aerosols in the atmosphere. Science 326, 1525–1529.

Kostiuk, L.W., Johnson, M.R., Thomas, G.P., 2004. University of Alberta flare research project final report, November 1996 – September 2004. Department of Mechanical Engineering, University of Alberta, Edmonton, Canada.

Lee, L., Wooldridge, P.J., de Gouw, J., Brown, S.S., Bates, T.S., Quinn, P.K., et al., 2015. Particulate organic nitrates observed in an oil and natural gas production region during wintertime. Atmos. Chem. Phys. 15, 9313–9325.

Massoli, P., Lambe, A.T., Ahern, A.T., Williams, L.R., Ehn, M., Mikkilä, J., et al., 2010. Relationship between aerosol oxidation level and hygroscopic properties of laboratory generated secondary organic aerosol (SOA) particles. Geophys. Res. Lett. 37, L24801. Available from: http://dx.doi.org/10.1029/2010GL045258.

Paerl, H.W., 1997. Coastal eutrophication and harmful algal blooms: importance of atmospheric deposition and groundwater as "new" nitrogen and other nutrient sources. Limnol. Oceanograp. 42, 1154–1165.

Prajapati, S.K., Tripathi, B.D., 2008. Biomonitoring seasonal variation of urban air Polycyclic Aromatic Hydrocarbons (PAHs) using *Ficus benghalensis* leaves. Environ. Pollut. 151, 543–548.

Riemer, N., West, M., 2013. Quantifying aerosol mixing state with entropy and diversity measures. Atmosp. Chem. Phys. 13, 11423–11439.

Riipinen, I., Yli-Juuti, T., Pierce, J.R., Tuukka Petaja, T., Worsnop, D.R., Kulmala, M., et al., 2012. The contribution of organics to atmospheric nanoparticle growth. Nat. Geosci. 5, 453–458.

Roy, A.A., Adams, P.J., Robinson, A.L., 2014. Air pollutant emissions from the development, production, and processing of Marcellus Shale natural gas. J. Air Waste Manage. Assoc. 64, 19–37.

Wang, J., Cubison, M.J., Aiken, A.C., Jimenez, J.L., Collins, D.R., 2010. The importance of aerosol mixing state and size-resolved composition on CCN concentration and the variation of the importance with atmospheric aging of aerosols. Atmos. Chem. Phys. 10, 7267–7283.

Zhang, Q., Alfarra, M.R., Worsnop, D.R., Allan, J.D., Coe, H., Canagaratna, M.R., et al., 2005. Deconvolution and quantification of hydrocarbon-like and oxygenated organic aerosols based on Aerosol Mass Spectrometry. Environ. Sci. Technol. 39, 4938–4952.

Chapter 6

Greenhouse Gas Emissions and Climate Impacts

Chapter Outline

THE PHYSICS OF CLIMATE CHANGE

The so-called "greenhouse effect" is a consequence of the laws of electromagnetic radiation and atomic physics. Light from the sun heats the earth, which then radiates photons in the infrared portion of the electromagnetic spectrum ($0.7-1000\ \mu m$). Instead of escaping directly to space, these photons are absorbed in the atmosphere by greenhouse gases—polyatomic molecules that rotate, vibrate, and bend faster as a consequence of absorbing infrared energy. In the denser lower atmosphere, the resulting excited state energy is transferred more quickly as heat by collisions with other molecules than by spontaneous re-radiation of photons to space, which would otherwise prevent excess warming from occurring (Houghton, 1977). In effect, greenhouse gases function as an invisible blanket that traps heat originally generated by the sun, preventing it from easily escaping the earth's atmosphere. Without them earth would be 33°C colder and in a frozen state (Hansen et al., 1981).

Water vapor is by far the most important terrestrial greenhouse gas, and is responsible for approximately 50% of the greenhouse effect, while CO$_2$ is responsible for roughly 20% (Schmidt et al., 2010). In the so-called "atmospheric window" between 8 and 13 μm, energy absorption by water vapor and CO$_2$ is weak and infrared photons more easily escape to space (see Fig. 6.1). Without the atmospheric window, the earth would become much too warm to support life. Outside this spectral region, infrared absorption is often "saturated," i.e. additional warming is inefficient due to the combination of strong absorption and high concentrations of the primary greenhouse gases (Goody and Yung, 1989). For this reason, species other than water vapor and CO$_2$ warm the earth most efficiently when they absorb radiation

Atmospheric Impacts of the Oil and Gas Industry. DOI: http://dx.doi.org/10.1016/B978-0-12-801883-5.00006-1

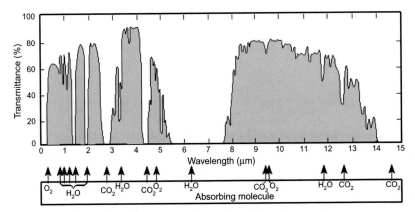

FIGURE 6.1   The transmission of radiation in the earth's atmosphere as a function of wavelength, and the main species that contribute to radiative absorption. *Wikipedia.*

in the relatively transparent atmospheric window. These include ozone and very long-lived (and therefore dangerous) man-made halocarbons such as sulfur hexafluoride (SF_6) and carbon tetrafluoride (CF_4).

The actual rise in average surface temperature due to greenhouse gases is not solely due to radiative processes, but also depends on dynamical and chemical feedbacks (National Research Council, 2003). These feedbacks (see Fig. 6.2) can amplify or damp the thermal response of the earth system to radiative forcing, defined as "the change in net (down minus up) irradiance (solar plus longwave; in W m^{-2}) at the tropopause" (Ramaswamy et al., 2001). An example of positive climate feedback is that due to atmospheric moisture content, which increases with temperature. Cloud feedback, on the other hand, can be either positive or negative depending on the type and altitude of cloud. High clouds trap more heat (positive feedback), whereas low clouds reflect more sunlight (negative feedback). Cloud feedback is complicated by the presence of aerosols, which may affect the physical and optical properties of clouds as well as contribute directly to radiative forcing. Besides the atmosphere itself, other components of the earth system contribute to feedback loops that affect climate, including the cryosphere (e.g., changes in the reflection of sunlight by the polar ice caps), the oceans (e.g., changes in the global storage and transport of heat), and the biosphere (e.g., changes in the global carbon cycle).

Predictions of climate change are ultimately dependent on the ability of numerical General Circulation Models (GCMs) to faithfully represent the complex feedbacks that determine the earth's response to changes in radiative forcing. The Intergovernmental Panel on Climate Change (IPCC) generates periodic summaries of the science behind global warming, including the evaluation of GCM predictions. As an aid to policy, the IPCC defines the equilibrium climate sensitivity as the change in global mean surface temperature at equilibrium that is caused by a doubling of the atmospheric CO_2 concentration.

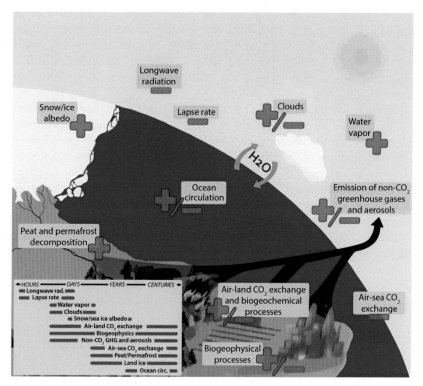

FIGURE 6.2 Illustration of climate feedbacks. *Fig. 1-02, Intergovernmental Panel on Climate Change (IPCC), 2013. Climate Change 2013: The Physical Science Basis. Contribution of Working Group I to the Fifth Assessment Report of the Intergovernmental Panel on Climate Change. In: Stocker, T.F., Qin, D., Plattner, G.-K., Tignor, M., Allen, S.K., Boschung, J., Nauels, A., Xia, Y., Bex, V., Midgley, P.M. (Eds.), Cambridge University Press, Cambridge, UK and New York, NY.*

In its latest assessment, the IPCC (2013) concluded with high confidence that equilibrium climate sensitivity is between 1.5°C and 4.5°C. The IPCC (2013) also judged that greenhouse gases likely contributed a historical global mean surface warming in the range of 0.5°C−1.3°C from 1951 to 2010.

GLOBAL WARMING POTENTIAL

From the standpoint of environmental policy, it is useful to compare the relative contributions of various greenhouse gases to climate change. This is done by means of an index known as Global Warming Potential (GWP), which compares the radiative forcing attributable to a given mass of a greenhouse gas to that of a similar mass of carbon dioxide. By definition, the GWP of CO_2 is 1. The GWP is a time-dependent index, and is usually calculated over time horizons of 20, 100 or 500 years. Its value depends not only on the infrared absorption spectrum of a greenhouse gas, but also on its atmospheric lifetime. A compound that strongly absorbs infrared radiation and has a long lifetime will have a large GWP, as is the case for CF_4, which has a lifetime of 50,000 years and a 100-year GWP of 7350 (IPCC, 2013). A compound with a short atmospheric lifetime will have a lower GWP with increasing time horizon.

Methane has a lifetime of 12.4 years (IPCC, 2013) and a strong infrared absorption band at around 7.5 μm at the very edge of the atmospheric window, in addition to one at around 3.3 μm. However, because of the relatively low concentration of methane compared to CO_2 (1.8 ppm by volume for CH_4 vs 400 ppm for CO_2), infrared absorption by methane is not saturated. For this reason, methane has a relatively high GWP of 84 over 20 years and 28 over 100 years, assuming no climate-carbon feedbacks (IPCC, 2013), in spite of the location of the methane bands.

Because of the methane's relatively high GWP, reductions in methane emissions are more effective in limiting the rise of global average surface temperature during this century than reductions in carbon dioxide emissions (see Fig. 6.3). This provides an enormous incentive to control leaks and other discharges of methane to the atmosphere from the oil and gas industry.

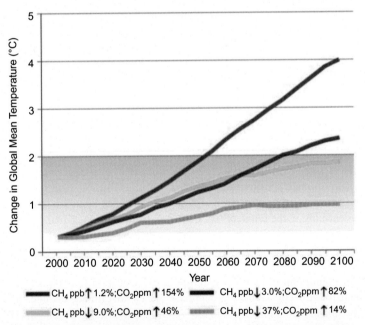

FIGURE 6.3 IPCC (2013) global warming projections for various emission scenarios and their accompanying changes in atmospheric concentrations of methane and carbon dioxide. *PSE Healthy Energy, 2015. Climate impacts of methane losses from modern natural gas and petroleum systems. Science Summary, November. Available at: <http://www.psehealthyenergy.org/data/ SS_Methane_Nov2015Final.pdf>.*

METHANE LEAKAGE

When burned efficiently by power plants, natural gas emits only half the amount of carbon as coal (IPCC, 2006). However, as pointed out by Howarth et al. (2011), the comparative advantage of natural gas relative to other fossil fuels may disappear when one considers emissions over its full lifecycle, beginning with the drilling of wells. Shale gas may be even worse than natural gas mined from conventional resources primarily due to methane released during well completion processes (Wang et al., 2014).

Wigley (2011) considered the complex atmospheric chemistry accompanying emissions of both greenhouse gases and aerosols from fossil fuel combustion, and concluded that unless methane leakage rates can be kept below 2%, shifting from coal to natural gas in electricity generation is ineffective in mitigating future climate change. Based on the most recent IPCC assessment, this critical leakage rate may be somewhat higher at 2.8% (PSE Healthy Energy, 2015). In a similar vein, substitution of compressed natural gas for diesel fuel in heavy-duty vehicles would require a methane emission rate of less than $\sim 1\%$ "from well-to-wheels" to ensure net climate benefits during this century (Alvarez et al., 2012; Camuzeaux et al., 2015; Tong et al., 2015).

A number of studies have used a top-down approach based on atmospheric measurements of hydrocarbon concentrations to estimate regional or global rates of methane leakage. Pétron et al. (2012) analyzed NOAA air samples in the Denver–Julesburg Basin of Colorado and concluded that on average, 4% (within a range of 2.3–7.7%) of all natural gas produced in the region was being leaked or vented to the atmosphere. Karion et al. (2013) derived a corresponding leakage rate of 6.2–11.7% from airborne measurements over the Uintah Basin, while Peischl et al. (2013) computed a 17% methane loss rate to the atmosphere from oil and gas activities in the Los Angeles Basin. Schweitzke et al. (2014) estimated natural gas fugitive emission rates using global atmospheric methane and ethane measurements over three decades, and inferred a most likely global average leakage rate of 2–4% since 2000. McKain et al. (2015) used a high-resolution transport model to estimate downstream natural gas emissions based on a year of continuous methane observations from four stations in the vicinity of Boston, Massachusetts. They estimated an average fractional loss rate to the atmosphere from natural gas transmission, distribution, and end use of 2.7% in the Boston region, more than twice as high as the fraction implied by a closely comparable emission inventory.

In contrast to approaches based on regional and global measurements, some studies have conducted site-specific measurements focusing on various aspects of the natural gas supply chain. Allen et al. (2013) measured methane emissions from completion flowbacks, equipment leaks, and pneumatic pumps and controllers at 190 onshore natural gas sites in the United States. Based on their computed emission factors, they estimated total annual emissions from the relevant source categories to be 957 Gg of methane, lower than the corresponding EPA national inventory estimate of ~ 1200 Gg. Combining EPA estimates for other categories with the emissions inferred from their measurements, Allen et al. (2013) estimated total methane emissions from natural gas production of 2300 Gg/year corresponding to a leakage rate of only 0.42%. Marchese et al. (2015) measured facility-level methane emissions at 114 natural gas gathering facilities and 16 processing plants in 13 US states. They then extrapolated these measurements to estimate nationwide emissions from gathering and processing operations based on facility counts obtained from state and national databases and probabilistic simulation. In doing so, they inferred total methane emissions of 2421 Gg/year for all US gathering and processing operations, implying a methane loss rate of 0.47% of 2012 gross production.

A possible explanation for the lower leakage rates reported by site-specific studies (which typically require industry cooperation) is that they do not represent the worst emitters, which may contribute a major fraction of total emissions even if they are a minority of facilities. Zimmerle et al. (2015) investigated methane emissions from the transmission and storage sector of the natural gas industry and found that "super-emitter" facilities constituted one of the largest source categories. Lavoie et al. (2015) analyzed airborne measurements during the 2013 Barnett Coordinated Campaign to quantify methane emission rates from eight different super-emitting point sources in the Barnett Shale. The inferred total emissions from these eight sources alone made up $\sim 9\%$ of the entire basin-wide emissions, and may have been easily missed by random sampling of facilities. Lyon et al. (2015) used measurements from the Barnett Coordinated Campaign as well as other sources of information to construct a spatially resolved methane emission inventory for the Barnett Shale that included high emission sites representing the very upper portion in the observed emissions distributions (see Fig. 6.4). These "fat-tail" emissions represented 19% of total emissions from shale mining activities, but only 2% of oil and gas sites. Taking super-emitters into account enabled Zavala-Araiza et al. (2015) to reconcile previously divergent top-down and bottom-up estimates of methane emissions from the Barnett Shale. The recent review of Allen (2016) provides further discussion of super-emitters in the context of air quality and climate change.

CO₂ FROM GLOBAL GAS FLARING

When barriers to natural gas markets and infrastructure exist, the associated natural gas that is a byproduct of oil extraction is often considered a nuisance rather than an economic resource. Associated gas not directly vented to the atmosphere is burned in flares, destroying most of its methane content, but emitting carbon dioxide instead. Elvidge et al. (2009) analyzed global satellite night light data and estimated that 2400 billion cubic meters (BCM)

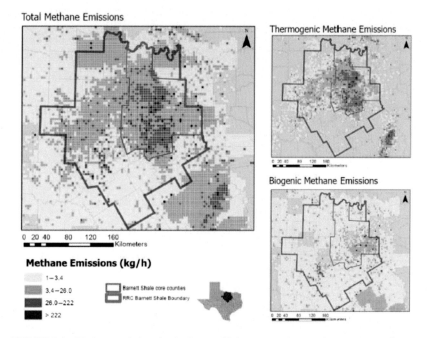

FIGURE 6.4 Methane emissions in the Barnett Shale. *Lyon, D.R., Zavala-Araiza, D., Alvarez, R.A., Harriss, R., Palacios, V., Lan, X., et al., 2015. Constructing a spatially resolved methane emission inventory for the Barnett Shale region. Environ. Sci. Technol. 49, 8147–8157.*

of natural gas was flared around the world between 1994 and 2008, roughly equivalent to 70% of US total greenhouse gas emissions in 2007. Elvidge et al. (2016) updated global gas flaring estimates for 2012 based on NASA/NOAA Visible Infrared Imaging Radiometer Suite (VIIRS) data. They computed a global total flared gas volume of 143 BCM, of which 129 BCM was from upstream flaring. Fig. 6.5 shows the top twenty countries that contributed to upstream flaring in 2012.

To mitigate the contribution of flaring to global climate change, the World Bank established the Global Gas Flaring Reduction (GGFR) Partnership in 2002. The GGFR is a public–private initiative that includes major oil companies, national and regional governments, and international institutions. The mission of the GGFR is to increase the use of associated natural gas by removing technical and regulatory barriers to flaring reduction, conducting research, disseminating best practices, and developing country-specific gas flaring reduction programs. In addition to the GGFR, the World Bank introduced the "Zero Routine Flaring by 2030" Initiative to eliminate routine flaring no later than 2030. To date, this initiative has been endorsed by 18 countries, including the governments of the United States, Canada and Russia.

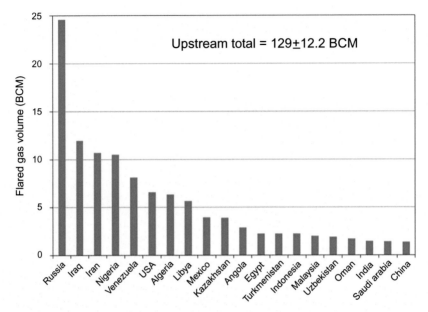

FIGURE 6.5   *Top 20 countries for upstream gas flaring. Elvidge, C.D., Zhizhin, M., Baugh, K., Hsu, F.-C., Ghosh, T., 2016. Methods for global survey of natural gas flaring from visible infrared imaging radiometer suite data. Energies 9 (1), 14. doi:10.3390/en9010014.*

TROPOSPHERIC OZONE

Ozone in the troposphere is a significant net absorber of infrared radiation, and may therefore contribute to climate change. Because of its relatively short lifetime of about 23 days (Young et al., 2013), the spatial distribution of tropospheric ozone is far from uniform. Nevertheless, it contributes approximately 0.4 W/m^2 to global radiative forcing compared to 2.8 W/m^2 for the well-mixed greenhouse gases (CO_2, methane, nitrous oxide, and halocarbons), according to the IPCC (2013) assessment.

On a global scale, methane and carbon monoxide are significant precursors to tropospheric ozone in addition to NO_x and VOCs. Since 1830, the mean tropospheric ozone burden has increased by roughly 30% in response to increases in anthropogenic emissions of all ozone precursors (Young et al., 2013). The oil and gas industry emits large quantities of longer-lived ozone precursors that may significantly contribute to a rise in the tropospheric ozone burden. For example, ethane is the second most abundant atmospheric hydrocarbon and a significant component of natural gas. Kort et al. (2016) analyzed 2014 airborne measurements collected over the Bakken Shale in North Dakota, and inferred ethane emissions of 0.23 Tg/year, equivalent to 1−3% of total global sources. This large and relatively recent ethane source at a single location illustrates the potential of widespread shale mining to significantly contribute to a rise in radiative forcing by tropospheric ozone.

REFERENCES

Allen, D.T., 2016. Emissions from oil and gas operations in the United States and their air quality implications. J. Air Waste Manage. Assoc. 66, 549–575.

Allen, D.T., Torres, V.M., Thomas, J., Sullivan, D.W., Harrison, M., Hendler, A., et al., 2013. Measurements of methane emissions at natural gas production sites in the United States. Proc. Natl. Acad. Sci. USA. 110, 17768–17773.

Alvarez, R.A., Pacala, S.W., Winebrake, J.J., Chameides, W.L., Hamburg, S.P., 2012. Greater focus needed on methane leakage from natural gas infrastructure. Proc. Natl. Acad. Sci. USA. 109, 6435–6440.

Camuzeaux, J.R., Alvarez, R.A., Brooks, S.A., Browne, J.B., Sterner, T., 2015. Influence of methane emissions and vehicle efficiency on the climate implications of heavy-duty natural gas trucks. Environ. Sci. Technol. 49, 6402–6410.

Elvidge, C.D., Zhizhin, M., Baugh, K., Hsu, F.-C., Ghosh, T., 2016. Methods for global survey of natural gas flaring from visible infrared imaging radiometer suite data. Energies 9 (1), 14. Available from: http://dx.doi.org/10.3390/en9010014.

Elvidge, C.D., Ziskin, D., Baugh, K.E., Tuttle, B.T., Ghosh, T., Pack, D.W., et al., 2009. A fifteen year record of global natural gas flaring derived from satellite data. Energies 2, 595–622.

Goody, R.M., Yung, Y.L., 1989. Atmospheric Radiation: Theoretical Basis. Oxford University Press, New York, NY, p. 544.

Hansen, J., Johnson, D., Lacis, A., Lebedeff, S., Lee, P., Rind, D., et al., 1981. Climate impact of increasing carbon dioxide. Science 213, 957–966.

Houghton, J.T., 1977. The Physics of Atmospheres. Cambridge University Press, Cambridge, UK and New York, NY, p. 203.

Howarth, R.W., Santoro, R., Ingraffea, A., 2011. Methane and the greenhouse-gas footprint of natural gas from shale formations. Clim. Change 106, 679. Available from: http://dx.doi.org/10.1007/s10584-011-0061-5.

Intergovernmental Panel on Climate Change (IPCC), 2006. IPCC guidelines for national greenhouse gas inventories. Geneva, Switzerland.

Intergovernmental Panel on Climate Change (IPCC), 2013. In: Stocker, T.F., Qin, D., Plattner, G.-K., Tignor, M., Allen, S.K., Boschung, J., Nauels, A., Xia, Y., Bex, V., Midgley, P.M. (Eds.), *Climate Change 2013: The Physical Science Basis. Contribution of Working Group I to the Fifth Assessment Report of the Intergovernmental Panel on Climate Change*. Cambridge University Press, Cambridge, UK and New York, NY.

Karion, A., Sweeney, C., Pétron, G., Frost, G., Hardesty, R.M., Kofler, J., et al., 2013. Methane emissions estimate from airborne measurements over a western United States natural gas field. Geophys. Res. Lett. 40, 4393–4397.

Kort, E.A., Smith, M.L., Murray, L.T., Gvakharia, A., Brandt, A.R., Peischl, J., et al., 2016. Fugitive emissions from the Bakken shale illustrate role of shale production in global ethane shift. Geophys. Res. Lett. 43, 4617–4623. Available from: http://dx.doi.org/10.1002/, 2016GL068703.

Lavoie, T.N., Shepson, P.B., Cambaliza, M.O.L., Stirm, B.H., Karion, A., Sweeney, A.C., et al., 2015. Aircraft-based measurements of point source methane emissions in the Barnett Shale basin. Environ. Sci. Technol. 49, 7904–7913.

Lyon, D.R., Zavala-Araiza, D., Alvarez, R.A., Harriss, R., Palacios, V., Lan, X., et al., 2015. Constructing a spatially resolved methane emission inventory for the Barnett Shale region. Environ. Sci. Technol. 49, 8147–8157.

Marchese, A.J., Vaughn, T.L., Zimmerle, D.J., Martinez, D.M., Williams, L.L., Robinson, A.L., et al., 2015. Methane emissions from United States natural gas gathering and processing. Environ. Sci. Technol. 49, 10718–10727.

McKain, K., Down, A., Raciti, S.M., Budney, J., Hutyra, L.R., Floerchinger, C., et al., 2015. Methane emissions from natural gas infrastructure and use in the urban region of Boston, Massachusetts. Proc. Natl. Acad. Sci. USA. 112, 1941−1946.

National Research Council (NRC), 2003. Understanding Climate Change Feedbacks. National Academies Press, Washington, DC, p. 166.

Peischl, J., Ryerson, T.B., Brioude, J., Aikin, K.C., Andrews, A.E., Atlas, E., et al., 2013. Quantifying sources of methane using light alkanes in the Los Angeles basin, California. J.Geophys. Res. Atmos. 118, 4974−4990.

Pétron, G., Frost, G., Miller, B.R., Hirsch, A.I., Montzka, S.A., Karion, A., et al., 2012. Hydrocarbon emissions characterization in the Colorado Front Range: a pilot study. J. Geophys. Res. Atmos. 117, D04303. Available from: http://dx.doi.org/10.1029/2011JD016360.

PSE Healthy Energy, 2015. Climate impacts of methane losses from modern natural gas and petroleum systems. Science Summary, November. Available at: <http://www.psehealthyenergy.org/data/SS_Methane_Nov2015Final.pdf>.

Ramaswamy, V., et al., 2001. Radiative forcing of climate change. In: Houghton, J.T., et al., (Eds.), Climate Change 2001: The Scientific Basis. Contribution of Working Group I to the Third Assessment Report of the Intergovernmental Panel on Climate Change. Cambridge University Press, Cambridge, UK and New York, NY, pp. 349−416.

Schmidt, G.A., Ruedy, R., Miller, R.L., Lacis, A.A., 2010. The attribution of the present-day total greenhouse effect. J. Geophys. Res. 115, D20106. Available from: http://dx.doi.org/10.1029/2010JD014287.

Schweitzke, S., Griffin, W.M., Magtthews, H.S., Bruhwiler, L.M.P., 2014. Natural gas fugitive emissions rates constrained by global atmospheric methane and ethane. Environ. Sci. Technol. 48, 7714−7722.

Tong, F., Jaramillo, P., Azevedo, I.M.L., 2015. Comparison of life cycle greenhouse gases from natural gas pathways for medium and heavy-duty vehicles. Environ. Sci. Technol. 49, 7123−7133.

Wang, Q., Chen, X., Awadhesh, N.J., Rogers, H., 2014. Natural gas from shale formation − the evolution, evidences and challenges of shale gas revolution in United States. Renew. Sustain. Energy Rev. 30, 1−28.

Wigley, T.M.L., 2011. Coal to gas: the influence of methane leakage. Clim. Change 108. Available from: http://dx.doi.org/10.1007/s10584-011-0217-3.

Young, P.J., et al., 2013. Pre-industrial to end 21st century projections of tropospheric ozone from the Atmospheric Chemistry and Climate Model Intercomparison Project (ACCMIP). Atmos. Chem. Phys. 13, 2063−2090.

Zavala-Araiza, D., Lyon, D.R., Alvarez, R.A., Davis, K.J., Harriss, R., Herndon, S.C., et al., 2015. Reconciling divergent estimates of oil and gas methane emissions. Proc. Natl. Acad. Sci. USA. 112, 15597−15602.

Zimmerle, D.J., Williams, L.L., Vaughn, T.L., Quinn, C., Subramanian, R., Duggan, G.P., et al., 2015. Methane emissions from the natural gas transmission and storage system in the United States. Environ. Sci. Technol. 49, 9374−9383.

Part II

Tools for Crafting Solutions

Chapter 7

Emission Inventories

Chapter Outline

IMPORTANCE OF EMISSIONS QUANTIFICATION

We have already discussed the problem of apparent underestimation of emissions from the oil and gas industry. From the standpoint of minimizing environmental impacts due to the industry's operations, this is a highly unsatisfactory situation. Without accurate emissions measurements, it is difficult to say whether or not control strategies are effective in mitigating unwanted releases to the atmosphere. For example, the success of international efforts to limit global warming depends on our ability to verify and track reductions of carbon dioxide, methane, and other greenhouse gases claimed by signatories to climate change treaties.

The most basic tool in the mitigation of atmospheric impacts is the emission inventory, which documents the types of activities that lead to emissions, the identities of the pollutants emitted, the location and time period for which the emissions are estimated, and the method used to estimate the emissions. This last aspect is the focus of this chapter, which provides an overview and specific examples of "bottom-up" methods for building emission inventories relevant to oil and gas development. In subsequent chapters, we will discuss various advanced techniques which, taken together, may improve our knowledge of emissions beyond current practice.

Atmospheric Impacts of the Oil and Gas Industry. DOI: http://dx.doi.org/10.1016/B978-0-12-801883-5.00007-3

OVERVIEW OF CURRENT EMISSIONS ESTIMATION TECHNIQUES

Bottom-up approaches for estimating emissions depend on data collected at the source or at representative groups of sources. Accepted methods include the following (EPA, 1999):

- Source sampling
- Emission factors
- Computer models
- Engineering calculations

Source Sampling

When cost is not a major consideration, analytical sampling at the source is the preferred method for emissions estimation. A probe or other device is used to collect pollutant samples that are either sent to a laboratory or analyzed using a portable instrument. Common detection principles for portable analyzers are flame or photo-ionization, nondispersive infrared absorption, gas thermal conductivity measurements, and heat measured by hot wire or catalytic oxidizer detectors (EPA, 2014). Emission rates averaged over the sampling time period are determined from the measured pollutant concentration, the volumetric gas flow rate and/or correlation equations (EPA, 1995). Source sampling is generally conducted for specific conditions (e.g., normal or maximum load), and does not account for variability over time.

Emission Factors

Emission factors relate pollutant emission rates to the level of a source activity, and are typically based on source tests at one or more facilities. They are intended to reflect industry-wide averages rather than specific conditions at a single location. The US EPA (1991) provides emission factors for criteria and hazardous air pollutants in the AP-42 database. Emission factors are in many cases also available from industrial trade associations or from manufacturers of specific pieces of equipment.

Computer Models

Computer models based on physical and chemical principles are available for estimating emissions from specific oil and gas processes. These models generally require various inputs, including physical and chemical properties of relevant materials, operating conditions, and the emission source characteristics (e.g., physical dimensions). An example is the GRI-GLYCalc model for emissions from glycol dehydration (Rueter et al., 1995).

Engineering Calculations

Calculations based on process-specific information may sometimes be performed without the need for a computer model. However, these calculations may be less accurate than emission factors if simplifying assumptions are used in lieu of real data.

In subsequent sections, we provide specific examples of each of these four methods. The reader should consult EPA (1999) and other literature for more comprehensive examples of these techniques.

LEAK MEASUREMENT DURING THE FORT WORTH NATURAL GAS AIR QUALITY STUDY

In 2010, the City of Fort Worth, Texas commissioned Eastern Research Group, Inc. (ERG) to perform an air quality monitoring study focused on natural gas development within the city limits. ERG (2011) collected data from 375 well pads, eight compressor stations, a gas processing plant, a saltwater treatment facility, a drilling operation, a hydraulic fracturing operation, and a completion operation. Fugitive emissions were monitored using an optical gas imaging (OGI) camera, a toxic vapor analyzer (TVA), a Hi-Flow Sampler and stainless steel canisters. ERG followed EPA Method 21 procedures (US Code of Federal Regulations, Title 40, Part 60, Appendix A) to survey ten per cent of valves, connectors and the other components using the TVA, in addition to any equipment for which significant leaks were indicated by the OGI camera. The volumetric flow rate of leaks at or above 500 ppm by volume was measured using the Hi-Flow Sampler. Selected gas samples were collected in stainless steel canisters for off-line laboratory analysis using a gas chromatograph-mass spectrometer (GC-MS).

ERG estimated emissions of total organic compounds (TOC) from the surveyed sites of 20,818 tons per year, more than 75% of which came from well pads. Approximately 98% of these emissions were for pollutants with relatively low toxicities, although several pollutants with high toxicities (e.g., benzene) were also emitted. At a small subset of sites, there were indications of large emission events likely associated with malfunctioning equipment, such as open hatches or corroded roofs on storage tanks. Table 7.1 summarizes the average and maximum emissions estimated by ERG for various site types.

FLARE EMISSION FACTORS

Emission factors are translated into emissions by multiplication with an activity rate as follows:

$$E_x = EF_x \times Q \tag{7.1}$$

where:

E_x = Emissions of pollutant x
EF_x = Emission factor of pollutant x
Q = Activity or production rate.

In the case of combustion sources, emission factors may be based on the Btu firing rate (MMBtu/h) rather than volume of fuel fired (MMscf/h, where scf refers to standard cubic feet of gas). In this case, Q in Eq. (7.1) becomes the Btu firing rate, which is then calculated based on the following equation:

$$Q = V \times H \qquad (7.2)$$

where:

V = Volume of fuel fired
H = Heating value of the fuel.

An example of the application of Eqs. (7.1) and (7.2) may be derived from the CO and NO_x emission factors for industrial flaring. The AP-42 values of these factors derived from EPA (1991) are 0.37 lbs of CO/MMBtu and 0.068 lbs of NO_x/MMBtu. (Note: In April 2015, the EPA released updated emission factors for flares, but the updated NO_x emission factor may be too high for oil and natural gas wellhead flares.) A typical value for H recommended by Bar-Ilan et al. (2008) for basins of the Central States

TABLE 7.1 Average and Maximum Emission Rates by Site Type

Site Type	TOC (tons/year)		VOC (tons/year)		HAP (tons/year)	
	Average	Maximum	Average	Maximum	Average	Maximum
Well pad	16	445	0.07	8.6	0.02	2
Well pad w/ compressors	68	4433	2	22	0.9	8.8
Compressor station	99	276	17	43	10	25
Processing facility	1293	1293	80	80	47	47
Saltwater treatment	1.5	1.5	0.65	0.65	0.4	0.4

Source: ERG (2011).

Regional Air Partnership (CENRAP) is 1209 MMBtu/MMscf. If the gas processing rate (V) is assumed to be 0.0002 MMscf/h, then:

$$Q = 0.0002 \text{ MMscf/h} \times 1209 \text{ MMBtu/MMscf}$$
$$= 0.242 \text{ MMBtu/h}$$
$$E_{CO} = 0.37 \text{ lbs/MMBtu} \times 0.242 \text{ MMBtu/h}$$
$$= 0.0895 \text{ lbs/h}$$
$$E_{NOx} = 0.068 \text{ lbs/MMBtu} \times 0.242 \text{ MMBtu/h}$$
$$= 0.0165 \text{ lbs/h}.$$

If we further assume that the flare operates for 8760 per year, then the annual CO and NO_x emissions from flaring are 784 and 145 lbs, respectively. Since 2000 lbs = 1 ton, these are equivalent to emission rates of 0.39 tons of CO/year and 0.07 tons of NO_x/year.

A STORAGE TANK EMISSIONS MODEL

For the purpose of generating regional emission inventories, direct sampling and analysis of fugitive releases from individual storage tanks at oil and gas sites is economically infeasible. A computer-based simulation model is therefore desirable to estimate the relevant emissions to the atmosphere from working, standing (breathing), and flashing losses. For this purpose, the American Petroleum Institute (API) developed the E&P TANK model (API, 1996–2014), which predicts methane, VOC, and HAP (BTEX and n-hexane) emissions from petroleum production field storage tanks, and is especially suitable for upstream operations.

E&P TANK is a steady-state model that requires site-specific information to determine emission rates. It quantifies flash emissions according to thermodynamic principles using the Peng and Robinson (1976) equation of state, assuming all components and phases are in equilibrium. Working and standing losses, on the other hand, are simulated differently according to the nature of the tank, and are represented either by a distillation column operation or a modified AP-42 method. The minimum inputs needed for the model are:

- separator oil composition;
- separator temperature and pressure;
- sales oil API gravity and Reid Vapor Pressure;
- sales oil production rate; and
- ambient temperature and pressure.

In addition, E&P TANK can accept detailed information about tank size, shape, and internal temperature to more precisely quantify emissions. A detailed sampling and analysis protocol is also provided for separator oil, since the composition of this oil is a key input to the model.

BLOWDOWN EMISSIONS ESTIMATED FROM THE DISPLACEMENT EQUATION

An example of a simple engineering calculation is the application of a displacement equation for estimating methane, VOC, and HAP emissions from blowdown events. Assuming that no chemical conversion occurs, the following equations can be used to estimate total and speciated VOC emissions:

$$E_{VOC} = Q \times MW \times X_{VOC} \times 1/C \qquad (7.3)$$

$$E_x = E_{VOC} \times X_x \qquad (7.4)$$

where:

E_{VOC} = Emissions of total VOC
Q = Volumetric flow rate/volume of gas processed
MW = Molecular weight of gas
X_{VOC} = Mass fraction of VOC in gas
C = Molar volume of ideal gas
E_x = Emissions of pollutant x
X_x = Mass fraction of species x in VOC.

The following example is reproduced from EPA (1999) and calculates VOC and benzene emissions resulting from blowdown of a group of compressor engines. Blowdown is assumed to occur 4 times per year, with a total volume of gas vented per event of 150 scf, so that the total annual volume of gas is 600 scf/year. Furthermore, we assume that the molecular weight of the gas is 29.2 lb/lb-mole, the mass fraction of VOC in the gas is 0.3, and the benzene content of VOC is 25% by weight. Therefore:

$Q = 600 \, \text{scf/year}$
$MW = 29.2 \, \text{lb/lb-mole}$
$X_{VOC} = 0.3 \, \text{lb VOC/lb}$
$C = 379 \, \text{scf/lb-mole@60}°\text{F, 1 atm}$
$E_{VOC} = 600 \, \text{scf/year} \times 29.2 \, \text{lb/lb-mole} \times 0.3 \, \text{lb VOC/lb} \times \text{lb-mole}/379 \, \text{scf}$
$\quad = 13.9 \, \text{lb VOC/year}$
$\quad = 13.9 \, \text{lb VOC/year} \times \text{ton}/2,000 \, \text{lb}$
$\quad = 0.007 \, \text{ton VOC/year}$
$X_{benzene} = 0.25 \, \text{lb benzene/lb VOC}$
$E_{benzene} = 0.25 \, \text{lb benzene/lb VOC} \times 13.9 \, \text{lb VOC/year}$
$\quad = 3.5 \, \text{lb benzene/year}$
$\quad = 3.5 \, \text{lb benzene/year} \times \text{ton}/2000 \, \text{lb}$
$\quad = 0.002 \, \text{ton benzene/year}.$

EXTRAPOLATION TO THE BASIN SCALE

The estimation methods presented in preceding sections apply to specific sites or pieces of equipment. To build a regional emission inventory, these estimates must be scaled up to the basin level. Aggregation of comprehensive site-specific data is impractical; thus surrogate data must be used to plausibly extrapolate the limited data available. A database of basin-wide statistics must therefore be compiled as a preliminary step towards building the regional inventory. For the upstream oil and gas industry, this database should include the location and number of active wells, their completion dates, the total number of drilling events during the time period of interest (i.e., the spud count), the total amount of gas production, and the total amount of oil or condensate production. In addition, it is very useful to conduct surveys in order to know the total population of each type of equipment within a given basin. These equipment types include drill rigs, compressor engines, artificial lift engines (pump jacks), heaters, oil or condensate tanks, and pneumatic devices. It is also important to include estimates of leak, completion venting, and blowdown emissions per well.

Bar-Ilan et al. (2008) provided recommendations on improving oil and gas emission inventories for the CENRAP states, including methods for scaling up emission estimates to the basin level. We will use wellhead compressor engines and flares as examples of how surrogate data may be used to perform the necessary extrapolation, based on their recommendations.

Wellhead Compressors

This example shows the utility of equipment surveys and well counts in extrapolating emissions from wellhead compressor engines. Emissions from individual wellhead compressors may be computed from the following equation:

$$E_{\text{engine}} = \frac{EF_i \times \text{HP} \times \text{LF} \times t_{\text{annual}}}{907,185} \tag{7.5}$$

where:

E_{engine} = emissions from a rich-burn or lean-burn compressor engine (ton/year/engine)
EF_i = emission factor of pollutant i (g/hp-h)
HP = engine horsepower (hp)
LF = engine load factor
t_{annual} = annual number of hours the engine is used (h/year).

A basin may be represented by a single rich- and lean-burn wellhead compressor engine make/model. Emissions are then scaled according to the following equation:

$$E_{engine,TOTAL} = (C_{Rich}E_{engine,Rich} + C_{Lean}E_{engine,Lean}) \times W_{TOTAL} \times f_{wellhead}$$

$$(7.6)$$

where:

$E_{engine,TOTAL}$ = total emissions from compressor engines in the basin (tons/year)

$E_{engine,Rich}$ = total emissions from a single representative rich-burn compressor engine (tons/year)

$E_{engine,Lean}$ = total emissions from a single representative lean-burn compressor engine (tons/year)

C_{Rich} = fraction of wellhead compressors in the basin that are rich-burn

C_{Lean} = fraction of wellhead compressors in the basin that are lean-burn

W_{TOTAL} = total well count in the basin

$f_{wellhead}$ = fraction of all wells in the basin with wellhead compressor engines.

Flaring

This example shows the utility of well production data and spud counts in extrapolating emissions for flares. Emissions from flaring depend on three separate process types according to the following equation:

$$E_{flare,TOTAL} = E_{flare,tank} + E_{flare,dehydration} + E_{flare,completion} \qquad (7.7)$$

where the expression on the left-hand side of Eq. (7.7) represents the total emissions due to flaring in the basin (tons/year), and the three terms on the right-hand side represent the separate contributions of flaring of stock tank flash gas, dehydration, and completion, respectively.

Emissions from flaring of oil and condensate stock tank flash gas are described by:

$$E_{flare,tank} = \left(\frac{EF_i \times Q_{flare,tank} \times H}{1000} \times \frac{P_{basin,liquid}}{1000} \right) / 2000 \qquad (7.8)$$

where:

EF_i = emission factor for pollutant i (lb/MMBtu)

$Q_{flare,tank}$ = volume of flash gas flared per unit of oil or condensate produced (MCF/1000 bbl)

H = local heating value of the gas (MMBTU/MMscf)

$P_{basin,liquid}$ = basin-wide oil or condensate production (bbl).

Note that the units MCF and bbl refer to million cubic feet of gas, and barrels of oil or condensate, respectively.

Emissions from flaring associated with dehydration processes are described by:

$$E_{\text{flare,dehydration}} = \left(\frac{EF_i \times Q_{\text{flare,dehydration}} \times H}{1000} \times \frac{P_{\text{basin,gas}}}{1,000,000} \right) / 2000 \quad (7.9)$$

where:

$Q_{\text{flare,dehydrator}}$ = dehydrator still vent gas flared per unit of gas produced (MCF/million MCF).

$P_{\text{basin,gas}}$ = basin-wide gas production (MCF).

Emissions from flaring associated with completion processes are described by:

$$E_{\text{flare,complexion}} = \left(\frac{EF_i \times Q_{\text{flare,complexion}} \times H}{1000} \times S_{\text{basin}} \right) / 2000 \quad (7.10)$$

where:

$Q_{\text{flare,completion}}$ = volume of completion venting gas flared per spud (MCF/spud)

S_{basin} = basin-wide spud count.

Growth and Control Factors

Emission inventories are usually compiled for a specific historical year. The resulting estimates, however, may need to be projected to future years. This is usually accomplished by multiplying the historical year emissions by a growth (or decline) factor based on a number of alternative sources, including:

- Broad regional oil and gas production outlooks (e.g., from the US EIA)
- Project-level estimates
- Published studies
- Extrapolation from historical data.

In addition, the existence of federal, state and local regulations must be taken into account, as well as the expected rule effectiveness. A control factor is typically used to reflect the impact of these legal mandates on projected emissions.

MIDSTREAM POINT SOURCE PERMITS

The preceding section applies primarily to the upstream sector of the oil and gas industry. For the midstream sector, emissions often come from facilities large enough to be officially designated as point sources, such as gas proces-sing plants and pipeline compressor stations. Limits to these emissions are typically stipulated in a legal permit. The USEPA requires midstream point sources to obtain air permits under two major programs mandated by the Clean Air Act.

New Source Review

The New Source Review (NSR) program requires industrial facilities to install the most effective pollution control equipment when they are built or when they are modified in a way that would cause a significant increase in emissions. Owners or operators of facilities are required to obtain permits that limit air emissions before they begin construction. Nonattainment NSR (NNSR) permits ensure that facilities in nonattainment areas do not worsen air where it is currently unhealthy to breathe. Permits in attainment areas are known as Prevention of Significant Deterioration (PSD) permits. They ensure that facilities do not degrade air that is currently clean. PSD and NNSR permits are often issued by state or local regulatory agencies.

Title V

Title V permits are legally enforceable documents issued to industrial facili-ties after they have begun to operate. These permits clarify what facilities must do to control air pollution to comply with either federal regulations or federally enforceable state regulations.

The emissions estimations required by air permits can be incorporated into emission inventories. For example, the USEPA National Emissions Inventory (NEI) accounts for emissions by midstream facilities based on data gathered by the states via permitting programs. The NEI is currently being expanded to include more upstream nonpoint sources, in addition to midstream point sources.

REFERENCES

American Petroleum Institute (API), 1996–2014. Production Tank Emissions Model, E&P TANK Version 3.0, User's Manual. API Publication 4697, Washington, DC.

Bar-Ilan, A., Parikh, R., Grant, J., Shah, T., Pollack, A.K., 2008. Recommendations for Improvements to the CENRAP States' Oil and Gas Emissions Inventories. Central States Regional Air Partnership, Oklahoma City, OK.

Eastern Research Group, Inc. (ERG), 2011. City of Fort Worth Natural Gas Air Quality Study. Fort Worth, TX.

Environmental Protection Agency (EPA), 1991. AP-42: compilation of air pollutant emission factors. Available at: <http://www.epa.gov/ttnchie1/ap42> (accessed 02.03.16.).

Environmental Protection Agency (EPA), 1995. Protocol for equipment leak emission estimates. EPA-453/R-95-017. Office of Air Quality Planning and Standards, Research Triangle Park, NC.

Environmental Protection Agency (EPA), 1999. Preferred and alternative methods for estimating air emissions from oil and gas field production and processing operations. In: EIIP Technical Report Series, Volume 2: Point Sources. Available at: <https://www.epa.gov/sites/production/files/2015-08/documents/ii10.pdf> (accessed 03.06.16.).

Environmental Protection Agency (EPA), 2014. Oil and natural gas sector leaks. Available at: <https://www3.epa.gov/airquality/oilandgas/pdfs/20140415leaks.pdf> (accessed 15.07.16.).

Peng, D.Y., Robinson, D.B., 1976. A new two-constant equation of state. Ind. Eng. Chem. Fundam. 15, 59−64.

Rueter, C.O., Reif, D.L., Myers, D.B., 1995. Glycol dehydrator BTEX and VOC emission testing results at two units in Texas and Louisiana. EPA/600/SR-96/046. Office of Research and Development, U.S. EPA, Research Triangle Park, NC.

Chapter 8

Ambient Air Monitoring and Remote Sensing

Chapter Outline

MONITORING APPLICATION TYPES

The science of environmental measurement is fast outstripping the capacity of regulatory agencies to absorb technology. This is partly due to the enormous investment that has gone into more established monitoring techniques, including the legal protocols that accompany their use in compliance and enforcement procedures, the extensive physical infrastructure required to implement them, and the lack of manpower willing and able to make use of newer methods. Nevertheless, there is a pressing need to displace older technologies to meet the demand for better information regarding the environmental impacts of the oil and gas industry and other major sources of pollution.

There are three applications of environmental measurement that are relevant to oil and gas development: (1) mapping and surveying, (2) baseline and trend monitoring, and (3) source attribution and emissions quantification. We will discuss each of these applications in turn.

Mapping and Surveying

An important aspect of monitoring is ensuring the proper operation of equipment on multiple sites, so that liabilities are minimized and production and profits are maximized. This implies the ability to account for spatial and

Atmospheric Impacts of the Oil and Gas Industry. DOI: http://dx.doi.org/10.1016/B978-0-12-801883-5.00008-5

temporal variability in emissions, such as might occur with pipeline leaks of methane. Mapping and surveying of equipment therefore requires a method for sampling and analysis that can be easily scaled up to larger areas, while delivering results quickly and efficiently so as to alert operators of existing problems in a timely fashion.

Baseline and Trend Monitoring

Baseline and trend monitoring addresses longer term issues compared to mapping and surveying. The general public needs assurance that oil and gas development does not significantly degrade the environment. Activists often demand that baseline monitoring be conducted prior to the onset of development in order to measure any subsequent changes in environmental quality that may be attributed to the oil and gas industry. Once the industry presence is established, trends in environmental parameters need to be measured for environmental quality to be maintained. Baseline and trend monitoring requires high precision (to better discriminate changes), repeatability of measurements, long term stability of instruments, and adequate spatial and temporal coverage of the affected areas.

Source Attribution and Emissions Quantification

It is necessary to quantify emissions from specific sources to determine whether or not environmental management strategies are successful, and to properly remediate problems when they are discovered. Accurate emission inventories are also required to assess environmental impacts through air quality modeling. Depending on the scale of the source attribution or emission inventory, this application requires moderate to very high spatial and temporal resolution, as well as an ability to measure correlations between various chemical species.

In the following sections, we will discuss different strategies for air sampling, recent developments in monitoring technology, and their relevance to the three application areas.

SAMPLING STRATEGIES

As previously discussed in Chapter 3, Toxic Air Pollution on Neighborhood Scales, conventional ambient air monitoring methods almost invariably suffer from the problem of spatial and temporal undersampling. In this section, we discuss a number of strategies that address this problem. This is followed by a survey of newer technologies that enable these strategies to be practically implemented.

Mobile Platforms With Real-Time Sensors

Mobile platforms equipped with fast, sensitive in situ sensors (with response times less than 10 s and detection limits around 1 ppb) can alleviate the undersampling problem by preferentially mapping ambient concentrations of pollutants in high emission areas at high time resolution, while still accurately measuring pollution gradients required for source attribution. Mobile platforms include automobiles, ships, manned or unmanned aircraft, and balloons. Herndon et al. (2005) provide examples of urban air quality measurements in Boston and Mexico City using a van-based mobile laboratory with a comprehensive suite of real-time instruments. Large mobile instrumentation suites are expensive to maintain and operate, and may not be feasible for longer-term applications such as monitoring trends. They are best used in intensive, short-term campaigns to discover problems, or to rigorously investigate issues flagged by other methods.

Instrumented mobile platforms may be deployed very effectively for source attribution and emissions quantification, especially when coupled with an inverse air quality model (see Chapter 9, Data Assimilation and Inverse Modeling). One of the first demonstrations of this combination of approaches was provided by Olaguer et al. (2013) based on mobile laboratory measurements outside a major refinery during the Formaldehyde and Olefins from Large Industrial Releases (FLAIR) sub-experiment of the 2009 SHARP campaign (see Chapter 4, Urban and Regional Ozone). The addition of real-time data broadcasting over the Internet enhances the power of mobile monitoring even further by enabling adaptive measurements to be made immediately in response to detected emission events. Fig. 8.1 shows a Web screenshot of a real time broadcast system as implemented by the Houston Advanced Research Center (HARC) during txhe BEE-TEX field study (Olaguer, 2015; see also Chapter 3, Toxic Air Pollution on Neighborhood Scales). This system allowed off-site personnel to direct mobile lab operations based on current observations, guide them towards areas with significant emission activities, and improve the quality of data assimilated by an inverse model.

The increasing use of Unmanned Aerial Vehicles (UAVs) or drones, together with the miniaturization of lasers and other environmental sensors, has opened up tremendous possibilities for the routine monitoring of large areas, with all the advantages of adaptive mobile sensing (see Fig. 8.2). However, regulations on the use of drones must be developed further in order to make environmental monitoring using drone-mounted sensors commonplace.

Remote Sensing

Remote sensing using passive or active optical techniques can be used to detect and/or measure air pollution while operating at a distance from sources. Techniques based on infrared (IR) radiation have shorter range compared to those based on ultraviolet (UV) or visible radiation.

FIGURE 8.1 Web screenshot of the HARC real-time broadcast system displaying mobile lab locations (*green dots*) and measurements (right hand side of screenshot). Concentration data are displayed for C2-benzenes (ethyl benzene + xylenes). *Olaguer, E.P., 2015. Overview of the Benzene and other Toxics Exposure (BEE-TEX) field study. Environ. Health Insights 9(S4), 1–6, http://dx.doi.org/10.4137/EHI.S15654.*

FIGURE 8.2 Hypothetical operation of an Unmanned Aerial Vehicle equipped with miniaturized sensors.

FIGURE 8.3 Infrared camera image of storage tank emissions. *US EPA.*

11/ 4/09 3.03.26PM   ⬜⬜⬜▷

An elementary application of remote sensing is the use of an optical gas imaging (OGI) camera to detect leaks and other emissions, as discussed in Chapter 7, Emission Inventories. The most common OGI cameras passively capture IR images of pollution plumes in real time based on the thermal contrast between the emitted gas and the background, but without quantifying the emissions from the source (see Fig. 8.3). More advanced versions can quantify hydrocarbon emissions and provide limited speciation based on pixel density and species response factors in a narrow wavelength range. Some commercially available hyper-cameras can quantify emissions for a broader range of compounds, but their ability to do this is limited by spectral line overlaps between many VOCs of interest and water vapor or carbon dioxide in the atmosphere. Despite their limitations, OGI cameras can play a useful role in mapping and surveying oil and gas sites by alerting operators of problems that may be investigated in greater depth with other techniques.

A second application of remote sensing is the measurement of ambient concentrations of pollutants averaged over an open path in the atmosphere. In passive remote sensing, the spectral absorption of light emitted by an industrial source or by the sun and/or sky over a path between the target emission source and a detector is measured and deconvoluted based on known absorption features to yield the path-averaged concentrations of the absorbing gases. If the wind speed is known, then the emission flux from the source can also be quantified. In active remote sensing, an artificial light source emits light towards a detector either: (1) at the opposite end of the light path (bi-static configuration), or (2) collocated with the light source (mono-static configuration) if the light is reflected by a mirror (i.e., a retro-reflector). Pollutant concentrations are then measured over the artificial light path.

While open path remote sensing with a single monitoring instrument yields only path-averaged concentrations, the combination of multiple, overlapping light paths from two or more geographically separated and/or positionally adjustable light sources (supplemented as necessary by several retro-reflectors) makes possible more detailed mapping of pollutant concentrations, including the identification and measurement of "hot spots," via computer aided tomography (CAT). The translation of path-averaged concentrations into a two-dimensional concentration map on the plane of the CAT scan measurements may be accomplished either with a purely mathematical method such as the Algebraic Reconstruction Technique (Gordon et al., 1970), or with a physical model supplemented by a data assimilation algorithm (Olaguer, 2011). Early applications of tomographic remote sensing in the atmosphere were performed by Todd et al. (2001) using Fourier Transform IR Spectroscopy (FTIR; Griffiths and de Haseth, 2007), and by Laepple et al. (2004) using UV Long Path (LP) Differential Optical Absorption Spectroscopy (DOAS; Platt and Stutz, 2008). More recently, LP-DOAS tomography was used to measure aromatic HAP concentrations near the Houston Ship Channel during the BEE-TEX field study (Olaguer, 2015; Stutz et al., 2016; Olaguer et al., 2016b).

The network of light paths used during BEE-TEX is illustrated in Fig. 8.4. Note that some of the intended light paths were inoperative due to

FIGURE 8.4 Tomographic network configuration in the Manchester neighborhood of Houston during the BEE-TEX study. *Red* lines indicate active light paths. *Olaguer, E.P., 2015. Overview of the Benzene and other Toxics Exposure (BEE-TEX) field study. Environ. Health Insights 9 (S4), pp. 1–6, http://dx.doi.org/10.4137/EHI.S15654.*

the blocking of sightlines by tall trees. Despite the degraded resolution of the CAT scan network, some interesting pollution plumes and emission sources were discovered during BEE-TEX that would otherwise have gone undetected, including nocturnal fugitive emissions from railcar loading and unloading operations (Olaguer et al., 2016b; Stutz et al., 2016).

Open path remote sensing is progressing to the point that networked instruments may in the near future be able to provide continuous long-term observations over entire neighborhoods, thereby making possible the measurement of pollution trends in a cost-effective manner. Tomographic measurements, in particular, facilitate inverse modeling (Olaguer, 2011), thus enabling source attribution and emissions quantification in addition to mapping and surveying of pollutants.

Distributed Sensor Networks

Mobile and remote sensing methods are generally perceived as expensive options, even though they may be cost-effective compared to more elementary techniques if labor and efficiency are factored in. An alternative to these methods is the use of large numbers of relatively inexpensive sensors linked by wireless or fiber optic networks. Environmentalists see this as an opportunity for citizen science in which nonexperts "crowd source" measurements on a large scale. However, the detection limits of current cheap sensors are usually not sufficiently low to make such citizen science an adequate source of monitoring information, except perhaps to flag very large emission events. There are also the issues of quality control and reliability when nonexperts are allowed to provide data, as well as the lack of suitable data interpretation engines to make sense of the measurements. A more promising application of distributed sensor networks is the facilitation of source sampling, if inexpensive sensors can be "pasted" onto industrial emission points, as this would not require low detection limits.

The South Coast Air Quality Management District (SCAQMD) of California has established an Air Quality Sensor Performance Evaluation Center (AQ-SPEC; see http://www.aqmd.gov/aq-spec) to inform citizens about commercially available low-cost sensors. The objectives of AQ-SPEC are to: (1) evaluate the performance of low-cost air quality sensors in both field and laboratory settings; (2) provide guidance for evolving sensor technology and data interpretation; and (3) catalyze the successful evolution, development and use of sensor technology.

ADVANCED MEASUREMENT TECHNOLOGIES

The measurement strategies discussed in the previous section depend on the availability of monitoring instruments that can deliver the required speed, sensitivity, and/or portability to accomplish the desired objectives.

We will examine various classes of instrument technologies based on their distinctive physical principles, with an emphasis on real-time detection and measurement of gaseous species. For a more comprehensive discussion of measurement technologies for aerosols, the reader is referred to the reviews of Canagaratna et al. (2007) and Amaral et al. (2015).

Optical Techniques

Optical techniques can be used to provide both in situ and remote sensing measurements. For in situ applications, an air sample is extracted from the atmosphere and enclosed within a gas cell through which a beam of light from a source is passed. The spectral absorption by constituent gases is then measured. The detection limit of the measurement can be reduced by lengthening the light path within the gas cell by means of a hollow waveguide (e.g., Harrington, 2000; Wilk et al., 2013), or by multiple reflections in a multipass cell or optical cavity. For example, the use of resonant optical cavities in Cavity Ring Down Spectroscopy (CRDS; Paul and Saykally, 1997) can achieve laser optical path lengths on the order of fifty kilometers.

Compact, robust and tunable lasers are now available that can target mid-IR ($2-25\,\mu m$) absorption spectra of one or more species. Semiconductor lasers include the tunable diode laser (TDL) and the quantum cascade laser (QCL). In diode lasers, a single photon is emitted as a result of a single electron-hole recombination. QCLs, on the other hand, produce a cascade of electrons down a series of quantum wells, enabling QCLs to emit several watts of peak power in pulsed operation and tens of milliwatts in continuous wave operation (which may yield higher sensitivity and precision). QCLs also provide a wider range of spectral coverage than TDLs, and can be tuned to particular wavelengths either by distributed feedback (DFB) through an additional Bragg grating, or by an external cavity which provides greater tuning range. Castillo et al. (2013) used a DFB QCL to perform simultaneous open path measurements of the greenhouse gases, methane and N_2O. Some QCLs are sensitive enough to provide high precision measurements of isotopic ratios (Eyer et al., 2016), which can be used to distinguish methane source signatures. In addition, tunable QCLs that have been miniaturized down to the size of coins and can be mounted on UAVs are now commercially available.

A third type of semiconductor laser is the interband cascade laser (ICL), which generates photons with interband transitions, rather than the intersubband transitions used in QCLs. ICLs require lower electrical input power than QCLs.

An alternative to direct laser absorption spectroscopy (LAS) is laser photo acoustic spectroscopy (LPAS). In this technique, the absorbed energy from a laser heats the target gas, thereby producing a measurable sound wave (Ball, 2006). Jahjah et al. (2014) used quartz-enhanced LPAS to achieve a methane detection limit of 13 ppb for a data acquisition time of 1 s.

Besides in situ measurement, lasers can also be used for active remote sensing. For example, portable active systems based on tunable diode LAS (TDLAS) technology are being used to detect methane leaks from oil and gas facilities at close range (tens of meters) as a faster, more sensitive alternative to passive OGI cameras. These instruments require a background target to reflect the light beam (e.g., a wall, grass, etc.).

A well-known active remote sensing technique is Light Detection and Ranging (LIDAR), wherein photons from a laser pulse that are backscattered by air molecules are collected by a detector and recorded as a function of time. The photon time-of-flight (ToF) is used to determine the range at which the scattering event occurred. Differential absorption LIDAR (DIAL) operates at two wavelengths, one on resonance and one off resonance of the molecular absorption of a target gas. The difference between the two signals is proportional to the gas concentration. DIAL can be used to measure emissions from industrial facilities, but because it is a short-range technique, access within the fence line is required. Robinson et al. (2011) applied the IR DIAL technique to measure VOC emissions from petrochemical facilities in the United Kingdom. They found that the measured emission fluxes exceeded conventional estimates by more than a factor of two.

Nonlaser light is also commonly used in remote sensing for air quality applications. For example, passive FTIR can be employed to measure flare emissions within industrial fence lines (URS, 2004). In this case, the hot flare gases act as the light source.

Another passive IR technique known as Solar Occultation Flux (SOF) makes use of the sun as a light source to measure pollutant column densities between the instrument and the sun, as well as corresponding fluxes (assuming that the wind speed is known). SOF is typically used outside industrial fence lines to measure pollutant fluxes both upwind and downwind of facilities, the net difference representing the facility emission flux. Mellqvist et al. (2010) deployed a van equipped with a SOF instrument to measure emissions of olefins in the Houston region during TexAQS II. They found discrepancies of an order of magnitude between their measured emission fluxes and corresponding emission inventory values, but agreement within 50% between SOF and airborne measurements. A limitation of the SOF technique is that it cannot be used to measure aromatics, which do not have significant IR absorption spectra.

We have already introduced the LP-DOAS technique in discussing sampling strategies. This method employs artificial UV and visible light sources such as Xenon arc lamps and Light Emitting Diodes (LEDs), the latter being a recent development (Stutz et al., 2016). There are also passive versions of DOAS that use the sun and sky as light sources. One of these is known as Imaging (I) DOAS because of its ability to image a combustion plume while also quantifying its emission flux (Pikelnaya et al., 2013). An I-DOAS measures the absorption of diffuse UV and visible sunlight to compute the amount of absorbing species between the instrument and the background atmosphere behind an emission plume. Emission fluxes are then inferred from absorber amounts based on the wind speed. Another passive version of DOAS is similar to the SOF technique, and is sometimes referred to as Sky DOAS if used to measure column densities and fluxes (e.g., Rivera et al., 2010), or as MultiAxis DOAS (MAX-DOAS) if used to measure areal fluxes of pollutants through a vertical cross section of the atmosphere.

Pikelnaya et al. (2013) used an I-DOAS to quantify emissions of formaldehyde from individual point sources in Texas City, Texas during the FLAIR experiment. They found that routine formaldehyde emissions from burning flares ranged from 0.3 to 2.5 kg/h. Stutz et al. (2011), on the other hand, deployed dual (upwind-downwind) stationary MAX-DOAS instruments during FLAIR. They found that net area-wide emission fluxes of primary HCHO from industrial facilities in Texas City were on the order of 20 kg/h.

Ionization Techniques

Mass spectrometry uses electromagnetic fields to guide charged particles, which are detected based on a mass-to-charge ratio. To be analyzed by a mass spectrometer, a chemical species must first be introduced in the gas phase into a vacuum and then ionized. The ionization process can either produce a molecular ion with the same molecular weight and elemental composition as the analyte, or it can produce fragment ions from smaller pieces of the analyte. Fragmentation makes identification of the parent compound more difficult, especially if multiple analytes produce similar ion fragments. For this reason, mass spectrometry (MS) is often combined with gas chromatography (GC), which separates analytes based on retention time in a capillary column prior to ionization and detection by a mass spectrometer. GC-MS, however, is a slow technique due to the time required for elution of analytes from the gas chromatograph, and is typically used for offline analysis of air samples collected in stainless steel canisters or adsorption cartridges.

The oldest MS ionization method is electron ionization, which employs an electron beam to strip neutral analytes of electrons, resulting in positively charged ions. Chemical ionization, on the other hand, uses a variety of reagent gas ions to selectively produce analyte ions. An advantage of the latter technique is that it produces much simpler mass spectra with reduced fragmentation compared to electron ionization. Chemical Ionization Mass Spectrometry (CIMS) has the advantage of speed compared to GC-MS, and is the basis for a variety of analytical instruments used in atmospheric research (e.g., Huey et al., 1995; Levy et al., 2014), including newer ones that employ faster and more sensitive ToF mass spectrometers compared to older quadrupole mass spectrometers (e.g., Yuan et al., 2016).

One of the most commonly used versions of the CIMS technique is Proton Transfer Reaction-Mass Spectrometry (PTR-MS; de Gouw and Warneke, 2007), which makes use of hydronium (H_3O^+) as a reagent to ionize air samples, although commercial instruments can also employ other reagent ions such as NO^+, O_2^+ and Kr^+. PTR-MS and related techniques played a key role during the BEE-TEX field study in the real time measurement of HAPs such as BTEX, styrene and 1,3-butadiene (Olaguer, 2015; Yacovitch et al., 2015; Olaguer et al., 2016a).

Warneke et al. (2014) conducted mobile PTR-Ms measurements of VOCs at or near oil and gas facilities in the Uinta Basin of Utah, including oil wells, evaporation ponds, compressor stations, and injection wells. They observed high mixing ratios of aromatics, alkanes, cycloalkanes, and methanol for both extended periods of time and during short-term spikes, which they attributed to local oil and gas sources. The mobile laboratory measurements confirmed results from an emission inventory, which identified the main VOC sources as dehydrators, oil and condensate tank flashing, and pneumatic devices and pumps.

Other Chemical Techniques

Traditional gas chromatography involves the separation of chemical compounds based on physical interaction with a specific column filling. A new analytical separation technique known as Membrane Inlet Mass Spectrometry (MIMS; Mach et al., 2015) has emerged that can deliver vastly improved speed compared to GC for field monitoring applications. MIMS employs semi-permeable membranes that selectively filter organic molecules and chemistries prior to analysis by a mass spectrometer. For example, polydimethylsiloxane (PDMS) membranes allow aromatic hydrocarbons such as BTEX to pass through while discriminating against aliphatics.

A chemical technique based on microfluidics may provide a new generation of inexpensive sensors with medium sensitivity. Microfluidic paper-based analytical devices (μPADs), originally developed for point-of-care medical diagnostics, are now being applied in environmental analyses (Meredith et al., 2016). A μPAD is a device in which very small (microliter) volumes of sample flow through a fibrous network by capillary action, guided by impermeable barriers. Ordinary paper can be cut to create flow channels that enable selective detection and measurement of analytes at microgram level by colorimetric reagents based on flow distance and gradients. Besides being very inexpensive, μPADs are portable and disposable, easy to make and use, and require minutes to analyze. For air quality, μPADs have mostly been employed as particulate sensors, but they can also be applied to gaseous pollutants, although VOC monitoring based on this technology is still in the very early stages of development.

REFERENCES

Amaral, S.S., de Carvalho Jr., J.A., Martins Costa, M.A., Pinheiro, C., 2015. An overview of particulate matter measurement instruments. Atmosphere 6, 1327—1345.

Ball, D.W., 2006. Photoacoustic spectroscopy. Spectroscopy 21, 14—16.

Canagaratna, M.R., Jayne, J.T., Jimenez, J.L., Allan, J.D., Alfarra, M.R., Zhang, Q., et al., 2007. Chemical and microphysical characterization of ambient aerosols with the Aerodyne aerosol mass spectrometer. Mass Spectrom. Rev. 26, 185—222.

Castillo, P.C., Sydoryk, I., Gross, B., Moshary, F., 2013. Ambient detection of CH_4 and N_2O by quantum cascade laser. In: Vo-Dinh, T., Lieberman, R.A., Gauglitz, G.G. (Eds.) Advanced Environmental, Chemical, and Biological Sensing Technologies X. Proceedings of SPIE, vol. 8718, 87180J (May 31, 2013). Available from: <http://dx.doi.org/10.1117/12.2016294>.

de Gouw, J.A., Warneke, C., 2007. Measurements of volatile organic compounds in the earth's atmosphere using proton-transfer reaction mass spectrometry. Mass Spectrom. Rev. 26, 223—257.

Eyer, S., Tuzson, B., Popa, M.E., van der Veen, C., Röckmann, T., Rothe, M., et al., 2016. Real-time analysis of $\delta^{13}C$- and δD-CH_4 in ambient air with laser spectroscopy: method development and first intercomparison results. Atmos. Meas. Tech. 9, 263—280.

Gordon, R., Bender, R., Herman, G.T., 1970. Algebraic reconstruction techniques (ART) for three-dimensional electron microscopy and X-ray photography. J. Theor. Biol. 29, 471—481.

Griffiths, P.R., de Haseth, J.A., 2007. Fourier Transform Infrared Spectrometry, 2nd ed Wiley, New York.

Harrington, J.A., 2000. A review of IR transmitting, hollow waveguides. Fiber Integrated Opt. 19, 211—217.

Herndon, S.C., Jayne, J.T., Zahniser, M.S., Worsnop, D.R., Knighton, B., Alwine, E., et al., 2005. Characterization of urban pollutant emission fluxes and ambient concentration distributions using a mobile laboratory with rapid response instrumentation. Farad. Discuss. 130, 327—339.

Huey, G.L., Hanson, D.R., Howard, C.J., 1995. Reactions of $SF6^-$ and I^- with atmospheric trace gases. J. Phys. Chem. 99, 5001—5008.

Jahjah, M., Ren, W., Stefański, P., Lewicki, R., Zhang, J., Jiang, W., et al., 2014. A compact QCL based methane and nitrous oxide sensor for environmental and medical applications. Analyst 139, 2065—2069.

Laepple, T., Knab, V., Mettendorf, K.-U., Pundt, I., 2004. Longpath DOAS tomography on a motorway exhaust gas plume: numerical studies and application to data from the BAB II campaign. Atmos. Chem. Phys. 4, 1323—1342.

Levy, M., Zhang, R., Zheng, J., Zhang, A.L., Xu, W., Gomez-Hernandez, M., et al., 2014. Measurements of nitrous acid (HONO) using ion drift-chemical ionization mass spectrometry during the 2009 SHARP field campaign. Atmos. Environ. 94, 231—240.

Mach, P.M., Wright, K.C., Verbeck, G.F., 2015. Development of multi-membrane near-infrared diode mass spectrometer for field analysis of aromatic hydrocarbons. J. Am. Soc. Mass Spectrom. 26, 281—285.

Mellqvist, J., Samuelsson, J., Johansson, J., Rivera, C., Lefer, B., Alvarez, S., et al., 2010. Measurements of industrial emissions of alkenes in Texas using the solar occultation flux method. J. Geophys. Res. Atmos. 115 (D7), D00K33. Available from: <http://dx.doi.org/10.1029/2008JD011682>.

Meredith, N.A., Quinn, C., Cate, D.M., Reilly, T.H., Volckens, J., Henry, C.S., 2016. Paper-based analytical devices for environmental analysis. Analyst 141, 1874—1887.

Olaguer, E.P., 2011. Adjoint model enhanced plume reconstruction from tomographic remote sensing measurements. Atmos. Environ. 45, 6980—6986.

Olaguer, E.P., 2015. Overview of the Benzene and other Toxics Exposure (BEE-TEX) field study. Environ. Health Insights 9 (S4), 1—6. Available from: <http://dx.doi.org/10.4137/EHI.S15654>.

Olaguer, E.P., Herndon, S.C., Buzcu Guven, B., Kolb, C.E., Brown, M.J., Cuclis, A.E., 2013. Attribution of primary formaldehyde and sulfur dioxide at Texas City during SHARP/ Formaldehyde and Olefins from Large Industrial Releases (FLAIR) using an adjoint chemistry transport model. J. Geophys. Res. Atmos. 118, 11,317−11,326.

Olaguer, E.P., Erickson, M.H., Wijesinghe, A., Neish, B.S., 2016a. Source attribution and quantification of benzene event emissions in a Houston Ship Channel community based on real time mobile monitoring of ambient air. J. Air Waste Manage. Assoc. 66, 164−172.

Olaguer, E.P., Stutz, J., Erickson, M.H., Hurlock, S.C., Cheung, R., Tsai, C., et al., 2016b. Real time measurement of transient event emissions of air toxics by tomographic remote sensing in tandem with mobile monitoring. Atmos. Environ. submitted for publication.

Paul, J.B., Saykally, R.J., 1997. Cavity ring down laser absorption spectroscopy. Anal. Chem. 69, 287A−292A.

Pikelnaya, O., Flynn, J., Tsai, C., Stutz, J., 2013. Imaging DOAS detection of primary formaldehyde and sulfur dioxide emissions from petrochemical flares. J. Geophys. Res. Atmos. 118, pp. 8716−8728.

Platt, U., Stutz, J., 2008. Differential Optical Absorption Spectroscopy: Principles and Applications. Springer, Berlin, p. 597.

Rivera, C., Mellqvist, J., Samuelsson, J., Lefer, B., Alvarez, S., Patel, M.R., 2010. Quantification of NO_2 and SO_2 emissions from the Houston Ship Channel and Texas City industrial areas during the 2006 Texas Air Quality Study. J. Geophys. Res. Atmos. 115, . Available from: <http://dx.doi.org/10.1029/2009JD012675>.

Robinson, R., Gardiner, T., Innocenti, F., Woods, P., Coleman, M., 2011. Infrared differential absorption Lidar (DIAL) measurements of hydrocarbon emissions. J. Environ. Monitor. 13, 2213−2220.

Stutz, J., Pikelnaya, O., Mount, G., Spinei, E., Herndon, S., Wood, E., et al., 2011. Quantification of hydrocarbon, NOx, and SO2 emissions from petrochemical facilities in Houston: interpretation of the 2009 FLAIR dataset, Report 10-045, Air Quality Research Program, University of Texas at Austin.

Stutz, J., Hurlock, S.C., Colosimo, S.F., Tsai, C., Cheung, R., Pikelnaya, O., et al., 2016. A novel dual-LED based long-path DOAS instrument for the measurement of aromatic hydrocarbons. Atmos. Environ., submitted for publication.

Todd, L.A., Ramanathan, M., Mottus, K., Katz, R., Dodson, A., Mihlan, G., 2001. Measuring chemical emissions using open-path fourier transform infrared (OP-FTIR) spectroscopy and computer-assisted tomography. Atmos. Environ. 35, 1937−1947.

URS Corporation, 2004. Passive FTIR Phase I testing of simulated and controlled flare systems. Texas Commission on Environmental Quality, Austin, TX.

Warneke, C., Geiger, F., Edwards, P.M., Dube, W., Pétron, G., Kofler, J., et al., 2014. Volatile organic compound emissions from the oil and natural gas industry in the Uintah Basin, Utah: oil and gas well pad emissions compared to ambient air composition. Atmos. Chem. Phys. 14, 10977−10988.

Wilk, A., Carter, J.C., Chrisp, M., Manuel, A.M., Mirkarimi, P., Alameda, J.B., et al., 2013. Substrate-integrated hollow waveguides: a new level of integration in mid-infrared gas sensing. Anal. Chem. 85, 11,205−11,210.

Yacovitch, T.I., Herndon, S.C., Roscioli, J.R., Floerchinger, C., Knighton, W.B., Kolb, C.E., 2015. Air pollutant mapping with a mobile laboratory during the BEE-TEX field study. Environ. Health Insights 9 (S4), 7−13. Available from: <http://dx.doi.org/10.4137/EHI.S15660>.

Yuan, B., Koss, A., Warneke, C., Gilman, J.B., Lerner, B.M., Stark, H., et al., 2016. A high-resolution time-of-flight chemical ionization mass spectrometer utilizing hydronium ions (H_3O^+ ToF-CIMS) for measurements of volatile organic compounds in the atmosphere. Atmos. Meas. Tech. 9, . Available from: <http://dx.doi.org/10.5194/amt-2016-21>.

Chapter 9

Data Assimilation and Inverse Modeling

Chapter Outline

THE EMERGENCE OF DATA ASSIMILATION

Modeling and monitoring in environmental science are undergoing a necessary evolution. Modeling applies mathematical and computational algorithms based on physical and chemical principles to the simulation of environmental phenomena. Monitoring, on the other hand, deploys various measurement techniques to determine pollutant concentrations and other environmental state variables. Although modeling and monitoring are complementary activities, they tend to be conducted in isolation from each other in traditional regulatory practice.

The emergence of data assimilation in the geophysical sciences suggests that modeling and monitoring should be highly coordinated. Models cannot rely entirely on first principles, since environmental phenomena are complex and often highly nonlinear. Moreover, the state of the environmental system is only partially known at any given time, so that a predictive model's initial conditions cannot be specified completely. The goal of data assimilation is to find the optimal balance between information derived from predictive models and information derived from observations. By constraining the behavior of models based on observations, data assimilation compensates for the lack of knowledge of many process details and of initial conditions. The use of data assimilation in environmental modeling began with numerical weather prediction, a history of which is provided by Kalnay (2003). Data assimilation in atmospheric chemistry is much more recent, and is summarized by Bocquet et al. (2015).

In this chapter, we illustrate the role that data assimilation plays in air quality science by means of examples from micro-scale modeling of tracer

Atmospheric Impacts of the Oil and Gas Industry. DOI: http://dx.doi.org/10.1016/B978-0-12-801883-5.00009-7

transport, i.e. the simulation of wind-driven dispersion of pollutants in the absence of significant chemistry roughly within 10 km downwind of sources. The utility of this approach to oil and gas issues will become readily apparent in the context of source attribution and emissions quantification via inverse modeling.

AN EULERIAN FORWARD AND ADJOINT TRANSPORT MODEL

We will demonstrate the data assimilation technique using the 4D variational (4Dvar) method based on both forward (prediction) and adjoint (correction) models (Sandu et al., 2005; Zou et al., 1997). In 4Dvar, a model is constrained by observations made at multiple times rather than at a single time as in 3Dvar. Other methods of data assimilation include the Ensemble Kalman Filter, an extension of the Kalman Filter relying on a spread of model states across an ensemble of simulations, and Nudging, which relaxes the model state towards the observations using artificial tendency terms in the prognostic equations.

We will apply the 4Dvar method within a 3D Eulerian finite-difference grid model that ignores chemical reactions, and is therefore applicable to relatively long-lived species such as methane or benzene. The governing equation of the forward model is the transport equation:

$$\partial C / \partial t = - \nabla \cdot (\boldsymbol{u} C) + \nabla \cdot \boldsymbol{K} \nabla C + E \qquad (9.1)$$

where C is the field of predicted concentrations, $\boldsymbol{u}$ is the wind vector, $\boldsymbol{K}$ the diffusion tensor, and E the emissions.

The 4Dvar method minimizes a cost function J, defined as the weighted least squares difference between the spatial and temporal predictions of a forward model and observations over an assimilation window or time period, plus other terms as necessary. Besides the forward model's concentration predictions, the cost function may be based on the initial or boundary conditions, and internal parameters such as emission rates and diffusion coefficients. Inverse modeling refers to the determination of emissions according to their influence on the agreement between the predicted concentrations and the observations, not necessarily through formal data assimilation.

An example of a cost function is as follows:

$$J = \frac{1}{2}(E - E^B)^T Q^{-1}(E - E^B) + \frac{1}{2}(K_h - K_h^B)^T D^{-1}(K_h - K_h^B)$$
$$+ \frac{1}{2}(C^0 - C^B)^T B^{-1}(C^0 - C^B) + \frac{1}{2}\sum_{k=0}^{N}(H_k C^k - C_{\text{obs}}^k)^T R_k^{-1}(H_k C^k - C_{\text{obs}}^k)$$

$$(9.2)$$

In Eq. (9.2), C^k is a column vector with dimension equal to the number of grid cells M, and represents the discretized field of predicted concentrations at time step k, where k ranges from 0 to N. Symbols containing the superscript B

refer to "background" or *a priori* estimates of the corresponding quantities. For example, C^B is the column vector of background initial concentrations, that is the "first guess" initial conditions. The parameter K_h refers to the horizontal diffusion coefficient, which may be optimized along with the emissions based on the information provided by the model horizontal gradients. (Note that other elements of the diffusion tensor may be included in the cost function as well.)

The observations that are assimilated by the 4Dvar method may be a complicated function of the forward model's predicted concentrations. This function is linearized so that if L is the number of observations corresponding to the dimension of the column vector of observations, C_{obs}^k, then the observation operator H_k in Eq. (9.2) is a matrix of dimension $L \times M$ that maps the model predictions onto the space of observations. For example, in an air quality CAT scan C_{obs}^k are the remote sensing observations (in units of concentration times ray path length). The observation operator is then simply the matrix of discretized ray cell path lengths (Olaguer, 2011).

Although the 4Dvar method is a deterministic technique, the cost function may be assigned a probabilistic interpretation analogous to the Kalman Filter, in which estimated parameters have a Gaussian distribution centered on the corresponding background values or measurements. In this interpretation, the quantities Q, D, B, and R in Eq. (9.2) are the error covariances of their associated parameters, and are related to the corresponding errors ε by Eq. (9.3):

$$
\begin{aligned}
Q &= <\varepsilon^E \varepsilon^{ET}> \\
D &= <\varepsilon^K \varepsilon^{KT}> \\
B &= <\varepsilon^B \varepsilon^{BT}> \\
R &= <\varepsilon^{obs} \varepsilon^{obsT}>.
\end{aligned}
\tag{9.3}
$$

To minimize the cost function, a Tangent Linear Model (TLM) is first constructed based on the forward model. The TLM in this case is:

$$
\frac{\partial \delta C}{\partial t} = -\nabla \cdot (u \delta C) + \nabla \cdot K \nabla \delta C + \delta E.
\tag{9.4}
$$

From the TLM an adjoint model is constructed governing the Lagrange multiplier λ, by which the constraint of the physical model is enforced, and which can be interpreted as the sensitivity of the cost function to the forward model concentration predictions. The continuous form of the adjoint model derived from the calculus of variations is given by (Sandu et al., 2005):

$$
-\frac{\partial \lambda}{\partial t} = \nabla \cdot (u \lambda) + \nabla \cdot K \nabla \lambda + \varphi
\tag{9.5}
$$

where φ is a forcing term related to the deviation of the model predictions from the observations, and the minus sign on the left hand side denotes backward integration in time. Note that in Eq. (9.5), advection is performed with the reverse wind. The continuous adjoint model is discretized prior to numerical integration. Diffusion is self-adjoint in both the continuous and discrete versions of the adjoint model. The discrete adjoint of advection can be obtained from automatic differentiation of the forward advection code, but the result does not necessarily commute with the continuous adjoint due to the numerical approximations involved.

To compute λ, the forward model is first run for the entire assimilation window, using a first guess solution as the initial conditions. At the end of the forward run, an adjoint final condition is then computed as input to the adjoint model as follows:

$$\lambda^N = H_N^T R_N^{-1}(H_N C^N - C_{obs}^N).$$ (9.6)

This information is then propagated backwards in time using the adjoint model, so that for $k = N - 1, \ldots 0$:

$$\lambda^k = \left(\frac{\partial C^{k+1}}{\partial C^k}\right)^T \lambda^{k+1} + H_k^T R_k^{-1}(H_k C^k - C_{obs}^k).$$ (9.7)

The corrections to optimized parameters are computed from λ as follows (Zou et al., 1997):

$$
\begin{aligned}
E &= E^B - Q\sum_{k=0}^{N-1}\left(\frac{\partial C^{k+1}}{\partial E(t^k)}\right)^T \lambda^{k+1} \\
K_h &= K_h^B - D\sum_{k=0}^{N-1}(\Delta t \nabla_h^2 C^k)^T \lambda^{k+1} \\
C^0 &= C^B - B\lambda^0
\end{aligned}
$$ (9.8)

where ∇_h^2 is the horizontal Laplacian.

Although the 4Dvar method, strictly interpreted as a deterministic approach, does not provide a formal update to the error covariances, one may appeal to a statistical interpretation as the basis for selecting their values. An important quantity in this regard is the Hessian of the cost function with respect to the optimized parameter P, which can be approximately linearized as follows:

$$\left(\frac{\partial^2 J}{\partial P^2}\right) \approx G^{-1} + \left(\frac{\partial C^N}{\partial P}\right)^T \left(\frac{\partial \lambda^N}{\partial C^N}\right)\left(\frac{\partial C^N}{\partial P}\right)$$ (9.9)

where G is the error covariance associated with P. The inverse of the Hessian corresponds to the updated error covariance in the statistical approach (Thacker, 1989).

Several cycles of forward and backward integration over the same assimilation window may be required for the cost function to converge. In each iteration cycle, the background values of the optimized parameters and their error covariances are replaced by their adjusted values from the previous iteration.

The end result of the 4Dvar method is an optimized set of initial/boundary concentrations and parameter values, including emissions from model grid points that have been assigned a nonzero initial estimate for the emission error covariance, which may be interpreted as a measure of emissions stochasticity. The larger this initial estimate is, the greater is the adjustment to the emissions that the adjoint model is predisposed to make at the corresponding grid point.

DEMONSTRATION OF THE ADJOINT METHOD IN AN OIL AND GAS APPLICATION

A forward and adjoint model with a formulation similar to that described above, but without optimization of the horizontal diffusion coefficient, was applied by Olaguer et al. (2016) to the measurement of BTEX emissions at a natural gas production facility in the Eagle Ford Shale. (This model was a simpler version of the HARC chemical transport model to be referred to in Chapter 10: Photochemical Simulation) Real-time measurements were obtained with a mobile laboratory equipped with a PTR-MS instrument, a portable meteorological station, and a Global Positioning System (GPS). Fig. 9.1 displays the benzene concentrations measured by the mobile lab on the day a major flare was observed to be in operation at the facility. The estimated average vector wind during the observation period was 3.3 m/s, 30.5° (NNE). This average background wind was extrapolated throughout the area of interest based a 3D digital morphological model of the production facility and the Quick Urban Industrial Complex (QUIC) model (Singh et al., 2008).

Olaguer et al. (2016) performed inverse modeling over a 700 m (x) × 700 m (y) × 200 m (z) domain covering the production facility and its immediate vicinity. The horizontal resolution employed for the simulation was 20 m, while 15 layers with increasing resolution towards the surface were deployed in the vertical direction. To ensure computational stability, the integration time step was set at 2 s. The transport model's numerical solvers included the Piecewise Parabolic Method for advection (Colella and Woodward, 1984), a semi-implicit scheme for vertical diffusion (Crank and Nicolson, 1947), and an explicit scheme for horizontal diffusion. The vertical turbulent diffusion coefficient was parameterized based on similarity theory, whereas the horizontal diffusivity was uniformly set at 50 m^2/s.

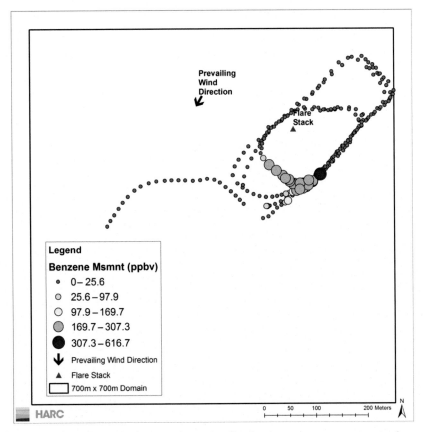

FIGURE 9.1   Mobile lab trajectory and corresponding benzene measurements at a natural gas production facility in the Eagle Ford Shale on the day a major flare was observed to be in operation. *Olaguer, E.P., Erickson, M.H., Wijesinghe, A., Neish, B.S., Williams, J., Colvin, J., 2016. Updated methods for assessing the impacts of nearby gas drilling and production on neighborhood air quality and human health. J. Air Waste Manage. Assoc. 66, 173–183.*

The observed large benzene plume was determined to be the result of a flare emission of 28 kg/hr averaged roughly over an hour, an order of magnitude greater than routine flare emissions measured at the same facility on other days. Corresponding emissions of toluene and C2-benzenes (ethylbenzene and xylenes) during the large flare event were 40 kg/h and 29 kg/h, respectively. Assuming that xylenes were the dominant component of C2-benzenes, the total masses of benzene, toluene and xylenes emitted by the flare in a single hour were 0.031, 0.044, and 0.032 tons, respectively, compared to annual limits of 0.11, 0.16, and 0.21 tons, respectively as stated in the facility permit for the flare. One may conclude, then, that major flaring only a few times per year may result in the technical violation of the facility permit.

DATA ASSIMILATION IN A GAUSSIAN DISPERSION MODEL

Due to the complexity of Eulerian grid models, the most common tools for simulating micro-scale pollutant transport are Gaussian plume dispersion models such as AERMOD (Cimorelli et al., 2004). These models rely on an analytical formula that assumes a stationary, nonzero homogeneous wind field and steady-state emissions. Lagrangian puff models such as the Second-order Closure Integrated Puff (SCIPUFF) model (Karamchandani et al., 2000) extend the applicability of the Gaussian method to nonstationary, nonhomogeneous conditions.

Gaussian puff models are based on the following equation:

$$c(x, y, z, t) = \frac{q_s}{\sqrt{2\pi}^3 \sigma_x \sigma_y \sigma_z} \exp\left[-\frac{(x - x_s - u(t - t_s))^2}{2\sigma_x^2} - \frac{(y - y_s - v(t - t_s))^2}{2\sigma_y^2} \right] \times$$
$$\left(\exp\left[-\frac{(z - z_s)^2}{2\sigma_z^2} \right] + \exp\left[-\frac{(z + z_s)^2}{2\sigma_z^2} \right] \right)$$

(9.10)

where $c(x, y, z, t)$ is the concentration field, q_s is the emission rate, t_s is the emission time, (x_s, y_s, z_s) are the source coordinates, (u, v) are the horizontal wind components, and $(\sigma_x, \sigma_y, \sigma_z)$ are the dispersion coefficients.

Albo et al. (2011) developed an inverse modeling system that combines an automatically-generated TLM of SCIPUFF with a cost function and a minimization algorithm that can search for multiple instantaneous or continuous sources without requiring an initial guess. Bieringer et al. (2015) developed a data assimilation methodology also based on SCIPUFF to improve the characterization of both emissions and winds near the surface for simulations of atmospheric transport and dispersion. In the method of Bieringer et al. (2015), an initial source estimate based on available observations of airborne contaminants and atmospheric conditions is first computed using a back-trajectory method. A variational refinement algorithm based on forward and adjoint versions of a simpler surrogate for the puff model is then used to iteratively refine the first guess source parameters and meteorological variables. Bieringer et al. (2015) applied their data assimilation algorithm to sensor observations during the Fusing Sensor Information from Observing Networks (FUSION) Field Trial 2007 (FFT07) chemical dispersion experiment (Storwold, 2007; Platt and DeRiggi, 2012). Table 9.1 shows their source estimation results when wind speed and direction were included among the optimized parameters.

TABLE 9.1 Source Estimation Results from Bieringer et al. (2015) for an FFT07 Trial When the Data Assimilation Algorithm Included Wind Speed and Direction Among the Optimized Parameters

	Source (kg)	Lat (deg)	Lon (deg)	Location Error (m)	Wind Speed (m/s)	Wind Direction (deg)
First guess	100	40.089678	112.969716	488	6.38	162.2
Estimate	3.8	40.091549	112.972652	173	2.31	173.3
Truth	1.158	40.0929425	112.973564	0	?	180

REFERENCES

Albo, S.E., Oluwole, O.O., Miyake-Lye, R.C., 2011. The Aerodyne Inverse Modeling System (AIMS): source estimation applied to the FFT 07 experiment and to simulated mobile sensor data. Atmos. Environ. 45, 6085–6092.

Bieringer, P.E., Rodriguez, L.M., Vandenberghe, F., Hurst, J.G., Bieberbach, Jr. G., Sykes, I., et al., 2015. Automated source term and wind parameter estimation for atmospheric transport and dispersion applications. Atmos. Environ. 122, 206–219.

Bocquet, M., Elbern, H., Eskes, H., Hirtl, M., Žabkar, R., Carmichael, G.R., et al., 2015. Data assimilation in atmospheric chemistry models: current statusand future prospects for coupled chemistry meteorology models. Atmos. Chem. Phys. 15, 5325–5358.

Cimorelli, A.J., Perry, S.G., Venkatram, A., Weil, J.C., Paine, R.J., Wilson, R.B., et al., 2004. AERMOD: description of model formulation. EPA-454/R-03-004. U.S. EPA, Research Triangle Park, NC.

Colella, P., Woodward, P.R., 1984. The Piecewise Parabolic Method (PPM) for gas-dynamical simulations. J. Comput. Phys. 54, 174–201.

Crank, J., Nicolson, P., 1947. A practical method for numerical evaluation of solutions of partial differential equations of the heat conduction type. Proc. Camb. Philol. Soc. 43, 50–67.

Kalnay, E., 2003. Atmospheric Modeling, Data Assimilation and Predictability. Cambridge University Press, Cambridge, UK.

Karamchandani, P., Santos, L., Sykes, I., Zhang, Y., Tonne, C., Seigneur, C., 2000. Development and evaluation of a state-of-the-science reactive plume model. Environ. Sci. Technol. 34, 870–880.

Olaguer, E.P., 2011. Adjoint model enhanced plume reconstruction from tomographic remote sensing measurements. Atmos. Environ. 45, 6980–6986.

Olaguer, E.P., Erickson, M.H., Wijesinghe, A., Neish, B.S., Williams, J., Colvin, J., 2016. Updated methods for assessing the impacts of nearby gas drilling and production on neighborhood air quality and human health. J. Air Waste Manage. Assoc. 66, 173–183.

Platt N., DeRiggi D., 2012. Comparative investigation of source term estimation algorithms for hazardous material atmospheric transport and dispersion prediction tools. IDA Doc D-4048, 4850 Mark Center Dr., Alexandria VA 22311.

Sandu, A., Daescu, D.N., Carmichael, G.R., Chai, T., 2005. Adjoint sensitivity analysis of regional air quality models. J. Comput. Phys. 204, 222–252.

Singh, B., Hansen, B.S., Brown, M.J., Pardyjak, E.R., 2008. Evaluation of the QUIC-URB fast response urban wind model for a cubical building array and wide building street canyon. Environ. Fluid Mech. 8, 281–312.

Storwold, D.P., 2007. Detailed test plan for the Fusing Sensor Information from Observing Networks (FUSION) Field Trial 2007 (FFT 07). West Desert Test Center, U.S. Army Dugway Proving Ground, WDTC Document No. WDTC-TP-07-078.

Thacker, W.C., 1989. The role of the Hessian matrix in fitting models to measurements. J. Geophys. Res. 94, 6177–6196.

Zou, X., Vandenberghe, F., Pondeca, M., Kuo, Y.-H., 1997. Introduction to adjoint techniques and the MM5 adjoint modeling system. NCAR/TN-435-STR. National Center for Atmospheric Research, Boulder, CO.

Chapter 10

Photochemical Simulation

Chapter Outline

ATMOSPHERIC TRANSFORMATIONS

The atmosphere functions as an active medium for environmental pollution. Chemical reactions in air continuously transform primary pollutants into numerous secondary by-products, beginning very near the source and spanning multiple scales from local to global. Without detailed knowledge of these transformations, it is very difficult to control harmful secondary pollutants such as ozone, or even to determine the ultimate fate of primary pollutants subject to regulation. In this chapter, we illustrate the complexity of atmospheric transformations in the specific context of near-source photochemistry.

Photochemical modeling has traditionally been conducted on urban, regional, and global scales. The realization that downstream petrochemical sources may sporadically release very large quantities of highly reactive species (Ryerson et al., 2003) has called attention to the rapid chemistry that may occur in the immediate vicinity of these sources. For example, Olaguer (2012b) used the HARC micro-scale chemical transport model (Olaguer, 2012a) to examine the near-source air quality impacts of large olefin flares in the Houston Ship Channel. (Note that the transport formulation of the HARC model was described in Chapter 9, Data Assimililation and Inverse Modeling.) Olaguer (2012b) found that a historical flare with a reported olefin emission rate exceeding 1400 lb/h (635 kg/h) may have resulted in a rate of increase in peak ozone greater than 40 ppb/h without any unusual meteorological conditions. Moreover, accompanying formaldehyde emissions from incomplete combustion in the flare may have enhanced peak ozone by as much as 16 ppb, and possibly contributed over 10 ppb to ambient formaldehyde. Olaguer (2013) employed an adjoint version of the full HARC model (including chemistry) to infer direct emissions of formaldehyde of 282 kg/h by an olefins flare system during a

Atmospheric Impacts of the Oil and Gas Industry. DOI: http://dx.doi.org/10.1016/B978-0-12-801883-5.00010-3

petrochemical facility shutdown, based on TexAQS II measurements of formaldehyde and other chemical species approximately 8 km downwind of the flare, where observed mixing ratios of formaldehyde were as high as 52 ppb.

We will elucidate the most important daytime gas phase reactions governing near-source atmospheric chemistry through the HARC chemical mechanism and solver (Olaguer, 2012a). In addition, the HARC chemical adjoint model (Olaguer, 2013) will be presented as a tool for inverse modeling of chemically reactive emissions. The mechanism version presented here has been updated with the latest available reaction rates and product yields. (Note that the olefin reaction numbering is slightly different from the previous version.)

THE HARC CHEMICAL MECHANISM

Air quality models employed in regulatory applications are based on chemical mechanisms that were designed for a wide range of scales beyond the urban environment. Unfortunately, this versatility comes at the expense of increasing numerical stiffness and computational cost. Unlike other mechanisms, the HARC chemical mechanism was designed to efficiently simulate rapid chemistry in and around fresh emission plumes under radical-limited (typically NO_x-rich) conditions. It provides a more detailed treatment of highly reactive VOCs such as olefins than conventional chemical mechanisms that lump these species together, while focusing on a smaller overall set of gas-phase reactions that can nonetheless successfully account for the budget of atmospheric radicals within current uncertainty.

Table 10.1 provides a list of abbreviations used for chemical species, while the 48 chemical reactions that make up the HARC mechanism are listed in Table 10.2A−G, along with the appropriate rate constants. Note that the use of parameterized photolysis rates in the HARC mechanism is easily overridden when a more sophisticated scheme such as TUV (Madronich and Flocke, 1998) or Fast-J (Wild et al., 2000) is available, but is probably adequate for many applications. Compared to the HARC mechanism, CB6 (Yarwood et al., 2010) has about 218 chemical reactions, while SAPRC07 (Carter, 2010) has about 260.

The treatment of olefins in the HARC mechanism is more schematic than that of the more comprehensive Master Chemical Mechanism (MCM; Saunders et al., 2003), but more detailed than in CB6, which lumps these compounds together. The condensed reaction schemes for the biogenic species, isoprene, and the aromatics, toluene and xylene, are truncated versions of those in CB6. Reaction products such as cresol have been omitted from the HARC mechanism, as we are interested in the immediate impact of isoprene and aromatics on the radical budget and ozone, rather than their influence on NO_x and HO_x ($=OH + HO_2$) through intermediate compounds further downwind of sources. The surrogate compound BVOC refers to "background" VOCs that are not explicitly represented in the HARC

TABLE 10.1 Chemical Abbreviations

Symbol	Definition
AO_2	Peroxy radical from OH reaction with propene, 1,3-butadiene and 1-butene
BO_2	Peroxy radical from OH reaction with 2-butene and isobutene
BUT1ENE	1-Butene
BUT2ENE	Cis-2-butene + trans-2-butene
BVOC	Volatile organic compounds not explicitly included in mechanism
IBUTENE	Isobutene
ISO_2	Peroxy radical from OH reaction with isoprene
ISOP	Isoprene
RNO_3	Organic nitrate
RO_2	$C_2H_5O_3 + AO_2 + BO_2 + ISO_2 + TO_2 + XO_2$
ROOH	Higher peroxide
TO_2	Peroxy radical from OH reaction with aromatics
TOL	Toluene
XO_2	Peroxy radical from OH reaction with background VOCs
XYL	Xylenes

TABLE 10.2A The HARC Chemical Mechanism *Photolysis and Thermal Decomposition*

Rate	Reaction	Rate Constant	Source
J_1	$NO_2 + h\nu \overset{O_2}{\rightarrow} NO + O_3$	$1.165 \times 10^{-2}(\cos\theta)^{0.244}$ $\times \exp\{-0.267 \sec\theta\}$	Saunders et al. (2003)
J_2	$O_3 + h\nu \rightarrow O(^1D) + O_2$	$6.073 \times 10^{-5}(\cos\theta)^{1.743}$ $\times \exp\{-0.474 \sec\theta\}$	Saunders et al. (2003)
J_3	$HCHO + h\nu \overset{O_2}{\rightarrow} CO + 2HO_2$	$4.642 \times 10^{-5}(\cos\theta)^{0.762}$ $\times \exp\{-0.353 \sec\theta\}$	Saunders et al. (2003)
J_4	$HCHO + h\nu \rightarrow CO + H_2$	$6.853 \times 10^{-5}(\cos\theta)^{0.477}$ $\times \exp\{-0.323 \sec\theta\}$	Saunders et al. (2003)
J_5	$HONO + h\nu \rightarrow OH + NO$	$2.644 \times 10^{-3}(\cos\theta)^{0.261}$ $\times \exp\{-0.288 \sec\theta\}$	Saunders et al. (2003)
J_6	$RNO_3 + h\nu \rightarrow HO_2 + NO_2$	$4.095 \times 10^{-6}(\cos\theta)^{1.111}$ $\times \exp\{-0.316 \sec\theta\}$	Saunders et al. (2003)
K_1	$HNO_4 \rightarrow HO_2 + NO_2$	$2.1 \times 10^{-27}\exp\left(\dfrac{10,900}{T}\right)$	JPL (2015)

Note: θ denotes the solar zenith angle. K denotes the equilibrium constant. J_6 is based on data for isopropyl nitrate. Reaction rate units are s^{-1}.

TABLE 10.2B The HARC Chemical Mechanism *Three-Body Reactions*

Rate	Reaction	Rate Constant	Source
l_1	$NO + OH \xrightarrow{M} HONO$	$k_0 = 7.0 \times 10^{-31} \left(\dfrac{300}{T}\right)^{2.6}$ $k_\infty = 3.6 \times 10^{-11} \left(\dfrac{300}{T}\right)^{0.1}$	JPL (2015)
l_2	$NO_2 + OH \xrightarrow{M} HNO_3$	$k_0 = 1.8 \times 10^{-30} \left(\dfrac{300}{T}\right)^{3.0}$ $k_\infty = 2.8 \times 10^{-11}$	JPL (2015)
l_3	$NO_2 + HO_2 \xrightarrow{M} HNO_4$	$k_0 = 1.9 \times 10^{-31} \left(\dfrac{300}{T}\right)^{3.4}$ $k_\infty = 4.0 \times 10^{-12} \left(\dfrac{300}{T}\right)^{0.3}$	JPL (2015)
l_4	$C_2H_4 + OH \xrightarrow{M} C_2H_5O_3$	$k_0 = 1.1 \times 10^{-28} \left(\dfrac{300}{T}\right)^{3.5}$ $k_\infty = 8.4 \times 10^{-12} \left(\dfrac{300}{T}\right)^{1.75}$	JPL (2015)
l_5	$C_3H_6 + OH \xrightarrow{M} AO_2$	$k_0 = 4.6 \times 10^{-27} \left(\dfrac{300}{T}\right)^{4.0}$ $k_\infty = 2.6 \times 10^{-11} \left(\dfrac{300}{T}\right)^{1.3}$	JPL (2015)

Note: $l([M], T) = \left\{ \dfrac{k_0[M]}{1 + \dfrac{k_0[M]}{k_\infty}} \right\} 0.6^{\left\{ 1 + \left[log_{10}\left(\frac{k_0[M]}{k_\infty}\right) \right]^2 \right\}^{-1}}$. Units are cm^6 molecule^{-2} s^{-1}.

TABLE 10.2C The HARC Chemical Mechanism $O(^1D)$ Reactions

Rate	Reaction	Rate Constant	Source
a_1	$H_2O + O(^1D) \rightarrow 2OH$	$1.63 \times 10^{-10} exp\left(\dfrac{60}{T}\right)$	JPL (2015)
a_2	$O(^1D) + N_2 \xrightarrow{O_2} O_3 + N_2$	$2.15 \times 10^{-11} exp\left(\dfrac{110}{T}\right)$	JPL (2015)
a_3	$O(^1D) + O_2 \rightarrow O_3$	$3.3 \times 10^{-11} exp\left(\dfrac{55}{T}\right)$	JPL (2015)

Note: Units for bimolecular reaction rates are cm^3 molecule^{-1} s^{-1}.

TABLE 10.2D The HARC Chemical Mechanism HO_x Reactions

Rate	Reaction	Rate Constant	Source
b_1	$O_3 + OH \rightarrow HO_2 + O_2$	$1.7 \times 10^{-12} \exp\left(\dfrac{-940}{T}\right)$	JPL (2015)
b_2	$O_3 + HO_2 \rightarrow OH + 2O_2$	$1.0 \times 10^{-14} \exp\left(\dfrac{-490}{T}\right)$	JPL (2015)
b_3	$NO + HO_2 \rightarrow NO_2 + OH$	$3.3 \times 10^{-12} \exp\left(\dfrac{270}{T}\right)$	JPL (2015)
b_4	$CO + OH \overset{O_2}{\rightarrow} HO_2 + CO_2$	$1.44 \times 10^{-13} +$ $3.43 \times 10^{-33}[M]$	Atkinson et al. (2006)
b_5	$HCHO + OH \overset{O_2}{\rightarrow} HO_2$ $+ CO + H_2O$	$5.5 \times 10^{-12} \exp\left(\dfrac{125}{T}\right)$	JPL (2015)
b_6	$BVOC + OH \rightarrow XO_2$	$b_6[BVOC] = r_{BVOC}$	Parameter
b_7	$HO_2 + OH \rightarrow H_2O + O_2$	$4.8 \times 10^{-11} \exp\left(\dfrac{250}{T}\right)$	JPL (2015)
b_8	$HO_2 + HO_2 \rightarrow H_2O_2 + O_2$	See note	JPL (2015)
b_9	$RO_2 + HO_2 \rightarrow ROOH$	$4.8 \times 10^{-13} \exp\left(\dfrac{800}{T}\right)$	CB6
b_{10}	$RNO_3 + OH \rightarrow 0.189HO_2 + NO_2$	7.2×10^{-12}	SAPRC07

Note: b_8 includes pressure and water vapor dependent terms, with the latter correction derived from JPL (2006), so that:

$$b_8 = \left\{ 3.0 \times 10^{-13} \exp\left(\frac{460}{T}\right) + 2.1 \times 10^{-33}[M] \exp\left(\frac{920}{T}\right) \right\}$$
$$\times \left\{ 1 + 1.4 \times 10^{-21}[H_2O] \exp\left(\frac{2200}{T}\right) \right\}$$

TABLE 10.2E The HARC Chemical Mechanism NO_x Reactions

c_1	$NO + O_3 \rightarrow NO_2 + O_2$	$3.0 \times 10^{-12} \exp\left(\dfrac{-1500}{T}\right)$	JPL (2015)
c_2	$HONO + OH \rightarrow NO_2 + H_2O$	$1.8 \times 10^{-11} \exp\left(\dfrac{-390}{T}\right)$	JPL (2015)
c_3	$HNO_4 + OH \rightarrow NO_2 + H_2O + O_2$	$1.3 \times 10^{-12} \exp\left(\dfrac{380}{T}\right)$	JPL (2015)

TABLE 10.2F The HARC Chemical Mechanism *Olefin Reactions*

Rate	Reaction	Rate Constant	Source
d_1	$C_2H_5O_3 + NO \rightarrow NO_2 + HO_2 + 2HCHO$	k_{RO_2NO}	See note
d_2	$AO_2 + NO \rightarrow NO_2 + HO_2 + HCHO$	k_{RO_2NO}	See note
d_3	$BO_2 + NO \rightarrow NO_2 + HO_2$	k_{RO_2NO}	See note
d_4	$C_4H_6 + OH \rightarrow AO_2$	$1.48 \times 10^{-11} \exp\left(\dfrac{448}{T}\right)$	MCMv3.3.1
d_5	$BUT1ENE + OH \rightarrow AO_2$	$6.6 \times 10^{-12} \exp\left(\dfrac{465}{T}\right)$	MCMv3.3.1
d_6	$BUT2ENE + OH \rightarrow BO_2$	$1.01 \times 10^{-11} \exp\left(\dfrac{550}{T}\right)$	MCMv3.3.1
d_7	$IBUTENE + OH \rightarrow BO_2$	$9.4 \times 10^{-12} \exp\left(\dfrac{505}{T}\right)$	MCMv3.3.1
d_8	$ISOP + OH \rightarrow ISO_2$	$2.7 \times 10^{-11} \exp\left(\dfrac{390}{T}\right)$	CB6
d_9	$ISO_2 + NO \rightarrow 0.883NO_2$ $+ 0.883HO_2 + 0.66HCHO$	k_{RO_2NO}	See note
d_{10}	$XO_2 + NO \rightarrow NO_2 + HO_2$	k_{RO_2NO}	See note
d_{11}	$C_2H_4 + O_3 \rightarrow 0.16OH + 0.16HO_2$ $+ HCHO + 0.51CO$	$9.1 \times 10^{-15} \exp\left(\dfrac{-2580}{T}\right)$	CB6
d_{12}	$C_3H_6 + O_3 \rightarrow 0.334OH + 0.23HO_2$ $+ 0.555HCHO + 0.378CO$	$5.5 \times 10^{-15} \exp\left(\dfrac{-1880}{T}\right)$	CB6
d_{13}	$C_4H_6 + O_3 \rightarrow 0.08OH + 0.42HO_2$ $+ 0.71HCHO + 0.63CO$	$1.34 \times 10^{-14} \exp\left(\dfrac{-2283}{T}\right)$	MCMv3.3.1
d_{14}	$BUT1ENE + O_3 \rightarrow 0.36OH + 0.28HO_2$ $+ 0.5HCHO + 0.18CO$	$3.55 \times 10^{-15} \exp\left(\dfrac{-1745}{T}\right)$	MCMv3.3.1
d_{15}	$BUT2ENE + O_3 \rightarrow 0.57OH + 0.125HO_2$ $+ 0.57CO$	$6.64 \times 10^{-15} \exp\left(\dfrac{-1059}{T}\right)$	MCMv3.3.1
d_{16}	$IBUTENE + O_3 \rightarrow 0.82OH + 0.41HO_2$ $+ 0.5HCHO$	$2.7 \times 10^{-15} \exp\left(\dfrac{-1630}{T}\right)$	MCMv3.3.1
d_{17}	$ISOP + O_3 \rightarrow 0.266OH + 0.066HO_2$ $+ 0.6HCHO + 0.066CO$	$1.0 \times 10^{-14} \exp\left(\dfrac{-1995}{T}\right)$	CB6

Note: $k_{RO_2NO} = 2.6 \times 10^{-12} \exp\left(\dfrac{365}{T}\right)$ is based on JPL (2015) for $C_2H_5O_2 + NO$. Yields for products of $RO_2 + NO$ are based roughly on MCMv3.3.1, except for $ISO_2 + NO$ and $TO_2 + NO$, for which product yields are based on CB6.

mechanism, including alkanes, acetaldehyde and higher aldehydes, and the isoprene decomposition products, methacrolein and methyl vinyl ketone. Reaction b_6 roughly accounts for these missing species, with the associated OH reactivity $r_{BVOC} = b_6[BVOC]$ treated as an external parameter.

TABLE 10.2G The HARC Chemical Mechanism *Aromatic Reactions*

Rate	Reaction	Rate Constant	Source
e_1	$TOL + OH \rightarrow 0.1OH + 0.25HO_2$ $+ 0.65TO_2$	$1.8 \times 10^{-12} \exp\left(\dfrac{340}{T}\right)$	CB6
e_2	$TO_2 + NO \rightarrow 0.86NO_2 + 0.86HO_2$	$k_{RO_2 NO}$	See note above
e_3	$XYL + OH \rightarrow 0.244OH + 0.213HO_2$ $+ 0.544TO_2$	1.85×10^{-11}	CB6

The HARC model predictions of HO_x, constrained by observations of longer-lived species, have been successfully benchmarked against measurements of the radical budget from the 2006 TexAQS II Radical and Aerosol Measurement Project (TRAMP; Lefer et al., 2010; Chen et al., 2010) for both high and low reactivity days (see Olaguer, 2012a, Supporting Information).

THE HARC CHEMICAL SOLVER

The numerical solver designed for the HARC mechanism takes advantage of the scale separation between fast chemical species, usually radicals, and slower species, including some members of the nitrogen oxide ($NO_x = NO + NO_2$) and odd oxygen [$O_x = O(^3P) + O(^1D) + O_3 + NO_2$] families. The particular chemical reaction determining this scale separation is $HO_2 + NO \rightarrow NO_2 + OH$ (reaction b_3). When NO is around a few ppb, the time constant for HO_2 loss is of the order of 1 s. The time constant for OH loss is an order of magnitude less than this, since the total OH reactivity in the presence of significant sources is typically over $10 \, s^{-1}$ (Mao et al., 2010). Thus, HO_x may be considered to be in chemical equilibrium near sources for time steps of the order of 10 s, which is adequate for wind speeds less than 10 m/s and horizontal grid lengths of around 200 m, based on the Courant-Friedrichs-Lewy criterion (Courant et al., 1928).

Assuming that the cycling between OH and HO_2 dominates the production and loss of HO_2 leads to the following expression for the ratio of HO_2 to OH:

$$\frac{[HO_2]}{[OH]} = \frac{b_1[O_3] + b_4[CO] + b_5[HCHO] + \zeta}{b_2[O_3] + b_3[NO]} \tag{10.1}$$

where the parameter ζ represents the total OH-to-HO_2 conversion rate associated with olefins (also referred to as alkenes), isoprene (a dialkene) and aromatic species:

$$\zeta = \epsilon + 0.225e_1[TOL] + 0.161e_3[XYL] \tag{10.2}$$

$$\epsilon = b_6[BVOC] + l_4[C_2H_4] + l_5[C_3H_6] + d_4[C_4H_6] + d_5[BUT1ENE]$$
$$+ d_6[BUT2ENE] + d_7[IBUTENE] + 0.883d_8[ISOP] + 0.503e_1[TOL]$$
$$+ 0.354e_3[XYL].$$

$$(10.3)$$

The parameter ϵ in Eq. (10.3) represents the rate at which OH is converted to HO_2 through intermediate reactions of RO_2 radicals with NO.

Assuming chemical equilibrium for HO_x, and combining the chemical terms for HO_x initiation and termination, yield a quadratic equation for OH:

$$A[OH]^2 + B[OH] - C = 0. \qquad (10.4)$$

The coefficients in Eq. (10.4) are given by the following expressions:

$$A = \left\{ 2b_7 + \left(\frac{\epsilon b_9}{k_{RO_2NO}[NO]} \right) + 2c_3K_1[NO_2] \right\} \frac{[HO_2]}{[OH]} + 2b_8 \left(\frac{[HO_2]}{[OH]} \right)^2 \quad (10.5)$$

$$B = l_1[NO] + l_2[NO_2] + 0.811b_{10}[RNO_3] + c_2[HONO] - (0.091e_1[TOL]$$
$$+ 0.079e_3[XYL])$$

$$(10.6)$$

$$C = 2a_1[H_2O][O(^1D)] + 2J_3[HCHO] + J_5[HONO] + J_6[RNO_3] + F[O_3]. \quad (10.7)$$

The first and last terms on the right hand side of Eq. (10.7) are computed from the equilibrium condition for $O(^1D)$, and rates of reaction between ozone and olefins:

$$\frac{[O(^1D)]}{[O_3]} = \frac{J_2}{a_1[H_2O] + a_2[N_2] + a_3[O_2]} \qquad (10.8)$$

$$F = 0.32d_{11}[C_2H_4] + 0.564d_{12}[C_3H_6] + 0.5d_{13}[C_4H_6] + 0.64d_{14}[BUT1ENE]$$
$$+ 0.695d_{15}[BUT2ENE] + 1.23d_{16}[IBUTENE] + 0.332d_{17}[ISOP].$$

$$(10.9)$$

The second term within the brackets in Eq. (10.5) represents the termination of HO_x due to peroxide formation, based on the assumption that the higher peroxy radicals referred to collectively as RO_2 are in chemical equilibrium predominantly due to fast termination by NO. The third term in brackets, on the other hand, represents the net termination of HO_x due to reactions involving pernitric acid (HNO_4), assuming that HNO_4 is in equilibrium.

While we may assume that HO_x is in chemical equilibrium under the conditions of interest to us, the same cannot be said of NO_x or ozone. Photolysis of NO_2, for example, occurs on timescales of the order of 100 s. Rather than resort to a complex chemical solver such as the SMVGEAR method (Jacobson and Turco, 1994) or a Rosenbrock solver (Sandu et al., 1997), the HARC model uses the efficient but highly accurate Euler Backward Iterative (EBI) scheme of Hertel et al. (1993) to predict the concentrations of NO, NO_2, and O_3. The EBI scheme uses analytical techniques for groups of strongly coupled species to solve a set of nonlinear chemical kinetic equations based on the implicit backward Euler method. The HARC chemical solver applies an EBI algorithm to the chemical group consisting of NO, NO_2, and O_3. Emission (E) and deposition (D) rates are added to chemical production and loss terms to update the concentrations of these species based on their values during the preceding time step, denoted by $[\ ]_0$.

The updates to the concentrations of NO and NO_2 are first computed from:

$$[NO] = ([NO]_1 - x)/\{1 + (D_{NO} + l_1[OH])\Delta t\} \tag{10.10}$$

$$[NO_2] = ([NO_2]_1 + x)/\{1 + (D_{NO_2} + l_2[OH])\Delta t\}, \tag{10.11}$$

where Δt is the time increment, and

$$[NO]_1 = [NO]_0 + E_{NO}\,\Delta t + (J_5[HONO] - \alpha)\Delta t \tag{10.12}$$

$$[NO_2]_1 = [NO_2]_0 + E_{NO_2}\Delta t + J_6[RNO_3]\,\Delta t + \alpha\Delta t + \{b_{10}[RNO_3] \\ + c_2[HONO]\}[OH]\Delta t \tag{10.13}$$

$$\alpha = \epsilon[OH]. \tag{10.14}$$

The parameter α is the rate of NO-to-NO_2 conversion resulting from VOC decomposition. The remaining inter-conversion term x is the root of a quadratic equation:

$$x = \frac{b - \sqrt{b^2 - 4ac}}{2a} \tag{10.15}$$

where,

$$a = 1 + \gamma \tag{10.16}$$

$$b = [O_3]_0 + (1 + 2\gamma)[NO]_1 + \lambda(1 + \gamma + \mu) \tag{10.17}$$

$$c = [NO]_1([O_3]_0 + \gamma[NO]_1) + \lambda(\gamma[NO]_1 - \mu[NO_2]_1) \tag{10.18}$$

$$\lambda = \frac{\{1 + (D_{NO} + l_1[OH])\Delta t\}(1 + \beta\Delta t)}{c_1\Delta t} \tag{10.19}$$

$$\mu = \frac{J_1\Delta t}{1 + (D_{NO_2} + l_2[OH])\Delta t} \tag{10.20}$$

$$\beta = D_{O_3} + a_1[H_2O]\frac{[O(^1D)]}{[O_3]} + b_1[OH] + b_2[HO_2] + G \tag{10.21}$$

$$G = d_{11}[C_2H_4] + d_{12}[C_3H_6] + d_{13}[C_4H_6] + d_{14}[BUT1ENE] + d_{15}[BUT2ENE] \\ + d_{16}[IBUTENE] + d_{17}[ISOP]$$

$$\tag{10.22}$$

$$\gamma = \frac{b_3[HO_2]\Delta t}{1 + (D_{NO} + l_1[OH])\Delta t}. \tag{10.23}$$

Once the updated concentrations of NO and NO_2 have been obtained, the concentration of ozone is then updated as follows:

$$[O_3] = \frac{([O_3]_0 + J_1[NO_2]\Delta t)}{\{1 + (c_1[NO] + \beta)\Delta t\}}. \tag{10.24}$$

The equilibrium concentrations for HO_x and the updates to NO, NO_2, and O_3 are computed iteratively based on the concentrations of longer-lived species from the previous time step. The results are then used to derive production and loss coefficients for 16 longer-lived species (including emission and deposition), whose concentrations are then updated using a simpler, noniterative backward Euler scheme. These 16 species are listed in Table 10.3, along with their chemical production and loss coefficients. In the case of organic nitrate, RNO_3, there

TABLE 10.3 Chemical Production and Loss (Plus Deposition) Coefficients of Transported Species

Index	Species	Production	Loss Coefficient
1	NO	$J_5[HONO] - \alpha - x/\Delta t$	$D_1 + l_1[OH]$
2	NO_2	$J_6[RNO_3] + \alpha + x/\Delta t + \{b_{10}[RNO_3] + c_2[HONO]\}[OH]$	$D_2 + l_2[OH]$
3	O_3	$J_1[NO_2]$	$c_1[NO] + \beta$
4	HONO	$l_1[NO][OH]$	$D_4 + J_5 + c_2[OH]$
5	HCHO	$\xi[OH] + \eta[O_3]$	$D_5 + J_3 + J_4 + b_5[OH]$
6	CO	$(J_3 + J_4)[HCHO] + b_5[HCHO][OH] + \theta[O_3]$	$D_6 + b_4[OH]$
7	C_2H_4		$D_7 + l_4[OH] + d_{11}[O_3]$
8	C_3H_6		$D_8 + l_5[OH] + d_{12}[O_3]$
9	C_4H_6		$D_9 + d_4[OH] + d_{13}[O_3]$
10	BUT1ENE		$D_{10} + d_5[OH] + d_{14}[O_3]$
11	BUT2ENE		$D_{11} + d_6[OH] + d_{15}[O_3]$
12	IBUTENE		$D_{12} + d_7[OH] + d_{16}[O_3]$
13	ISOP		$D_{13} + d_8[OH] + d_{17}[O_3]$
14	TOL		$D_{14} + e_1[OH]$
15	XYL		$D_{15} + e_3[OH]$
16	RNO_3		$D_{16} + J_6 + b_{10}[OH]$

is no local chemical production, and replenishment within the computational domain can only occur by inflow from the boundary. Organic nitrate formation outside the domain is assumed to be the main source of RNO_3, while its loss in the domain interior serves as a HO_x termination reaction.

Note that in the expressions for the chemical production of HCHO and CO, there appear the following coefficients:

$$\xi = 2l_4[C_2H_4] + l_5[C_3H_6] + d_4[C_4H_6] + d_5[\text{BUT1ENE}] + 0.66d_8[\text{ISOP}] \tag{10.25}$$

$$\eta = d_{11}[C_2H_4] + 0.555d_{12}[C_3H_6] + 0.71d_{13}[C_4H_6] + 0.5d_{14}[\text{BUT1ENE}] \\ + 0.5d_{16}[\text{IBUTENE}] + 0.6d_{17}[\text{ISOP}] \tag{10.26}$$

$$\theta = 0.51d_{11}[C_2H_4] + 0.378d_{12}[C_3H_6] + 0.63d_{13}[C_4H_6] + 0.18d_{14}[\text{BUT1ENE}] \\ + 0.57d_{15}[\text{BUT2ENE}] + 0.066d_{17}[\text{ISOP}]. \tag{10.27}$$

THE HARC CHEMICAL ADJOINT

The HARC adjoint model is based on the continuous adjoints of transport and chemistry, and does not rely on automatic differentiation of the forward model code. To compute the adjoint of the chemical model, we must first define the chemical Jacobian in terms of the production and loss coefficient of transported species i, denoted respectively by P_i and L_i (which includes the deposition term D_i), and its concentration C_i (molecules/cm^3):

$$\mathscr{J}_{ij} = \frac{\partial(P_i - L_iC_i)}{\partial C_j}. \tag{10.28}$$

The chemical adjoint is then simply the transpose of the chemical Jacobian multiplied by the adjoint vector with components λ_ι.

From the definition of the partial derivative, $\partial C_i / \partial C_j = \delta_{ij}$ (i.e., the Kronecker delta function). We can then derive the following analytical expressions from Eqs. (10.1) through (10.3):

$$\frac{\partial}{\partial C_j} \frac{[HO_2]}{[OH]} = \frac{b_1\delta_{3,j} + b_5\delta_{5,j} + b_4\delta_{6,j} + \frac{\partial \zeta}{\partial C_j}}{b_2[O_3] + b_3[NO]} - \frac{[HO_2]}{[OH]}\left(\frac{b_3\delta_{1,j} + b_2\delta_{3,j}}{b_2[O_3] + b_3[NO]}\right). \tag{10.29}$$

$$\frac{\partial \zeta}{\partial C_j} = \frac{\partial \epsilon}{\partial C_j} + 0.225e_1\delta_{14,j} + 0.161e_3\delta_{15,j}. \tag{10.30}$$

$$\frac{\partial \epsilon}{\partial C_j} = l_4 \delta_{7,j} + l_5 \delta_{8,j} + d_4 \delta_{9,j} + d_5 \delta_{10,j} + d_6 \delta_{11,j} + d_7 \delta_{12,j} + 0.883 d_8 \delta_{13,j}$$

$$+ 0.503 e_1 \delta_{14,j} + 0.354 e_3 \delta_{15,j}.$$

(10.31)

From Eqs. (10.4) through (10.9), we likewise derive:

$$\frac{\partial [OH]}{\partial C_j} = \left(\frac{1}{2A[OH] + B} \right) \left(\frac{\partial C}{\partial C_j} - [OH] \frac{\partial B}{\partial C_j} - [OH]^2 \frac{\partial A}{\partial C_j} \right)$$

(10.32)

$$\frac{\partial A}{\partial C_j} = \left\{ 2b_7 + \left(\frac{\epsilon b_9}{k_{RO_2NO}[NO]} \right) + 2c_3 K_1[NO_2] + 4b_8 \frac{[HO_2]}{[OH]} \right\} \frac{\partial}{\partial C_j} \frac{[HO_2]}{[OH]}$$

$$- \frac{[HO_2]}{[OH]} \left\{ \left(\frac{\epsilon b_9}{k_{RO_2NO}[NO]^2} \right) \delta_{1,j} - 2c_3 K_1 \delta_{2,j} - \left(\frac{b_9}{k_{RO_2NO}[NO]} \right) \frac{\partial \epsilon}{\partial C_j} \right\}$$

(10.33)

$$\frac{\partial B}{\partial C_j} = l_1 \delta_{1,j} + l_2 \delta_{2,j} + c_2 \delta_{4,j} - 0.091 e_1 \delta_{14,j} - 0.079 e_3 \delta_{15,j} + 0.811 b_{10} \delta_{16,j}$$

(10.34)

$$\frac{\partial C}{\partial C_j} = \left\{ F + 2a_1[H_2O] \frac{[O(^1D)]}{[O_3]} \right\} \delta_{3,j} + J_5 \delta_{4,j} + 2J_3 \delta_{5,j} + J_6 \delta_{16,j} + [O_3] \frac{\partial F}{\partial C_j}$$

(10.35)

$$\frac{\partial F}{\partial C_j} = 0.32 d_{11} \delta_{7,j} + 0.564 d_{12} \delta_{8,j} + 0.5 d_{13} \delta_{9,j} + 0.64 d_{14} \delta_{10,j} + 0.695 d_{15} \delta_{11,j}$$

$$+ 1.23 d_{16} \delta_{12,j} + 0.332 d_{17} \delta_{13,j}.$$

(10.36)

We will also need the following analytical expressions for various derivatives:

$$\frac{\partial \alpha}{\partial C_j} = [OH] \frac{\partial \epsilon}{\partial C_j} + \epsilon \frac{\partial [OH]}{\partial C_j}$$

(10.37)

$$\frac{\partial L_3}{\partial C_j} = c_1 \delta_{1,j} + \left(b_1 + b_2 \frac{[HO_2]}{[OH]} \right) \frac{\partial [OH]}{\partial C_j} + b_2 [OH] \frac{\partial}{\partial C_j} \frac{[HO_2]}{[OH]} + \frac{\partial G}{\partial C_j}$$ (10.38)

$$\frac{\partial G}{\partial C_j} = d_{11} \delta_{7,j} + d_{12} \delta_{8,j} + d_{13} \delta_{9,j} + d_{14} \delta_{10,j} + d_{15} \delta_{11,j} + d_{16} \delta_{12,j} + d_{17} \delta_{13,j}$$

(10.39)

$$\frac{\partial \xi}{\partial C_j} = 2l_4 \delta_{7,j} + l_5 \delta_{8,j} + d_4 \delta_{9,j} + d_5 \delta_{10,j} + 0.66 d_8 \delta_{13,j}$$ (10.40)

$$\frac{\partial \eta}{\partial C_j} = d_{11}\delta_{7,j} + 0.555d_{12}\delta_{8,j} + 0.71d_{13}\delta_{9,j} + 0.5d_{14}\delta_{10,j} + 0.5d_{16}\delta_{12,j} + 0.6d_{17}\delta_{13,j}$$

$$(10.41)$$

$$\frac{\partial \theta}{\partial C_j} = 0.51d_{11}\delta_{7,j} + 0.378d_{12}\delta_{8,j} + 0.63d_{13}\delta_{9,j} + 0.18d_{14}\delta_{10,j} + 0.57d_{15}\delta_{11,j}$$

$$+ 0.066d_{17}\delta_{13,j}.$$

$$(10.42)$$

The individual chemical Jacobian terms may now be evaluated, and are listed in Table 10.4.

Note that we have neglected the terms involving the derivative of x (see Eq. 10.15 and Table 10.3) in the Jacobian matrix. As the model time increment approaches zero,

$$x \to \Delta t\{c_1[NO][O_3] + b_3[HO_2][NO] - J_1[NO_2]\}. \qquad (10.43)$$

The nearly photostationary state close to NO_x sources causes the opposing terms on the right-hand side of Eq. (10.43) to be in approximate balance. Computational instability may then occur for small values of the dimensionless ratio, $x/(\alpha\Delta t)$. Ignoring the derivative of x avoids this computational instability at the expense of some inaccuracy in the partitioning (rather than the total amount) of NO_x by the adjoint model, which is partly compensated by the forward model.

The accuracy of the chemical adjoint may be checked by computing a matrix of adjoint sensitivities as follows:

$$\lambda_C = \frac{\partial C^N}{\partial C} \qquad (10.44)$$

where C is the vector of transported species concentrations and C^N is the final concentration vector. Olaguer (2013, see Supporting Information) computed adjoint sensitivities based on the original HARC chemical Jacobian and backwards integration using a semi-implicit scheme. He compared resulting values of λ_C^0 (where the superscript refers to the value of the adjoint sensitivity at the time $t = 0$) to "brute force" (BF) sensitivities of final concentrations to initial values derived by running the forward chemistry model. The adjoint and BF sensitivities were found to be in reasonable agreement despite the use of the continuous adjoint for the chemistry and the neglect of the Jacobian terms containing the derivative of x. Subsequent versions of the HARC model may restore these neglected terms whenever $x/(\alpha\Delta t)$ is above a threshold level.

TABLE 10.4 Chemical Jacobian Terms

$$\mathcal{J}_{1,j} = -(D_1 + l_1[\text{OH}])\delta_{1,j} + J_5\delta_{4,j} - l_1[\text{NO}]\frac{\partial[\text{OH}]}{\partial C_j} - \frac{\partial\alpha}{\partial C_j}$$

$$\mathcal{J}_{2,j} = -(D_2 + l_2[\text{OH}])\delta_{2,j} + c_2[\text{OH}]\delta_{4,j} + (J_6 + b_{10}[\text{OH}])\delta_{16,j} + (b_{10}[\text{RNO}_3] + c_2[\text{HONO}]$$
$$- l_2[\text{NO}_2])\frac{\partial[\text{OH}]}{\partial C_j} + \frac{\partial\alpha}{\partial C_j}$$

$$\mathcal{J}_{3,j} = J_1\delta_{2,j} - L_3\delta_{3,j} - [\text{O}_3]\frac{\partial L_3}{\partial C_j}$$

$$\mathcal{J}_{4,j} = l_1[\text{OH}]\delta_{1,j} - (D_4 + J_5 + c_2[\text{OH}])\delta_{4,j} + (l_1[\text{NO}] - c_2[\text{HONO}])\frac{\partial[\text{OH}]}{\partial C_j}$$

$$\mathcal{J}_{5,j} = \eta\delta_{3,j} - (D_5 + J_3 + J_4 + b_5[\text{OH}])\delta_{5,j} + [\text{O}_3]\frac{\partial\eta}{\partial C_j} + [\text{OH}]\frac{\partial\xi}{\partial C_j} + (\xi - b_5[\text{HCHO}])\frac{\partial[\text{OH}]}{\partial C_j}$$

$$\mathcal{J}_{6,j} = \theta\delta_{3,j} + (J_3 + J_4 + b_5[\text{OH}])\delta_{5,j} - (D_6 + b_4[\text{OH}])\delta_{6,j} + [\text{O}_3]\frac{\partial\theta}{\partial C_j} + (b_5[\text{HCHO}] - b_4[\text{CO}])\frac{\partial[\text{OH}]}{\partial C_j}$$

$$\mathcal{J}_{7,j} = -d_{11}[\text{C}_2\text{H}_4]\delta_{3,j} - (D_7 + l_4[\text{OH}] + d_{11}[\text{O}_3])\delta_{7,j} - l_4[\text{C}_2\text{H}_4]\frac{\partial[\text{OH}]}{\partial C_j}$$

$$\mathcal{J}_{8,j} = -d_{12}[\text{C}_3\text{H}_6]\delta_{3,j} - (D_8 + l_5[\text{OH}] + d_{12}[\text{O}_3])\delta_{8,j} - l_5[\text{C}_3\text{H}_6]\frac{\partial[\text{OH}]}{\partial C_j}$$

$$\mathcal{J}_{9,j} = -d_{13}[\text{C}_4\text{H}_6]\delta_{3,j} - (D_9 + d_4[\text{OH}] + d_{13}[\text{O}_3])\delta_{9,j} - d_4[\text{C}_4\text{H}_6]\frac{\partial[\text{OH}]}{\partial C_j}$$

$$\mathcal{J}_{10,j} = -d_{14}[\text{BUT1ENE}]\delta_{3,j} - (D_{10} + d_5[\text{OH}] + d_{14}[\text{O}_3])\delta_{10,j} - d_5[\text{BUT1ENE}]\frac{\partial[\text{OH}]}{\partial C_j}$$

$$\mathcal{J}_{11,j} = -d_{15}[\text{BUT2ENE}]\delta_{3,j} - (D_{11} + d_6[\text{OH}] + d_{15}[\text{O}_3])\delta_{11,j} - d_6[\text{BUT2ENE}]\frac{\partial[\text{OH}]}{\partial C_j}$$

$$\mathcal{J}_{12,j} = -d_{16}[\text{IBUTENE}]\delta_{3,j} - (D_{12} + d_7[\text{OH}] + d_{16}[\text{O}_3])\delta_{12,j} - d_7[\text{IBUTENE}]\frac{\partial[\text{OH}]}{\partial C_j}$$

$$\mathcal{J}_{13,j} = -d_{17}[\text{ISOP}]\delta_{3,j} - (D_{13} + d_8[\text{OH}] + d_{17}[\text{O}_3])\delta_{13,j} - d_8[\text{ISOP}]\frac{\partial[\text{OH}]}{\partial C_j}$$

$$\mathcal{J}_{14,j} = -(D_{14} + e_1[\text{OH}])\delta_{14,j} - e_1[\text{TOL}]\frac{\partial[\text{OH}]}{\partial C_j}$$

$$\mathcal{J}_{15,j} = -(D_{15} + e_3[\text{OH}])\delta_{15,j} - e_3[\text{XYL}]\frac{\partial[\text{OH}]}{\partial C_j}$$

$$\mathcal{J}_{16,j} = -(D_{16} + J_6 + b_{10}[\text{OH}])\delta_{16,j} - b_{10}[\text{RNO}_3]\frac{\partial[\text{OH}]}{\partial C_j}$$

INVERSE MODELING OF CHEMICALLY REACTIVE EMISSIONS

Olaguer et al. (2013) used the adjoint version of the HARC model with photochemistry to perform 4Dvar inverse modeling of an industrial release of formaldehyde (HCHO) and sulfur dioxide (SO_2) during the 2009 SHARP/FLAIR campaign in Texas City, Texas. (SO_2 was treated as a conservative

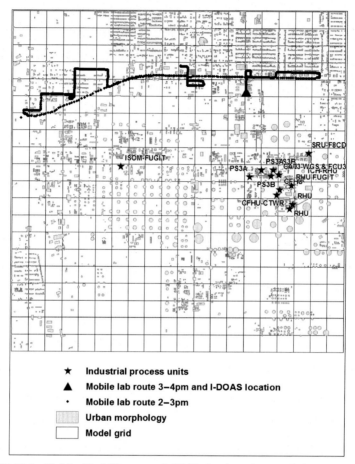

FIGURE 10.1   Model grid of Olaguer et al. (2013) showing urban morphology, approximate mobile lab route on May 13, 2009, and locations of major sources.

tracer in the HARC model, separate from the 16 species of Table 10.3.) The release was attributed to a Fluidized Catalytic Cracking Unit (FCCU) and desulfurization processes at one of the largest refineries in the United States. The source attribution was based on real-time quantum cascade laser and pulsed fluorescence measurements by a mobile laboratory, and a high resolution, 3D morphological model of the emitting petrochemical complex and surrounding urban canopy derived from LIDAR data.

Fig. 10.1 shows the model grid of Olaguer et al. (2013), along with the mobile lab route and main sources in the refinery. The modeling was performed on a 4 km × 4 km horizontal grid with 200 m resolution, and with winds extrapolated from meteorological measurements using the QUIC model (see Chapter 9, Data Assimilation and Inverse Modeling). The inverse model estimated that 18 kg/h of primary HCHO was emitted by the refinery during the incident. This estimate agreed very closely with simultaneous and independent remote sensing measurements based on both Imaging and Multi-Axis DOAS (see Chapter 8, Ambient Air Monitoring and Remote Sensing). The inferred HCHO-to-SO_2 molar emission ratio was similar to that computed directly from ambient air measurements during the release. In addition, the model-estimated HCHO-to-CO molar emission ratio for combustion units ranged from 2% to about 7%, consistent with other observationally-based estimates obtained during SHARP.

Olaguer and Lefer (2014) simulated wintertime ozone in the Uinta Basin of Utah by applying the HARC model on a 20 km × 20 km horizontal grid with 400 m resolution, constrained by ozone, reactive nitrogen, HCHO, and hydrocarbon measurements at the Horsepool site during the 2013 UBWOS campaign. The kinematic effects of topography were accounted for using the QUIC wind model based on assimilated meteorological measurements. In addition, a database of reported and/or permitted emissions from oil and gas sites was used to construct first guess estimates for the inverse model. Olaguer and Lefer (2014) successfully reproduced an observed rapid 3-h rise in ozone at the Horsepool site, and inferred primary HCHO emissions of 10 kg/h from oil and gas facilities in the Uinta Basin, similar to the output of a large refinery.

REFERENCES

Atkinson, R., Baulch, D.L., Cox, R.A., Crowley, J.N., Hampson, R.F., Hynes, R.G., et al., 2006. Evaluated kinetic and photochemical data for atmospheric chemistry: volume II − gas phase reactions of organic species. Atmos. Chem. Phys. 6, 3625−4055.

Carter WPL, 2010. Development of the SAPRC-07 chemical mechanism and updated ozone reactivity scales. Report to the California Air Resources Board, Contracts No. 03-318, 06-408, and 07-730, University of California, Riverside, CA.

Chen, S., Ren, X., Mao, J., Chen, Z., Brune, W.H., Lefer, B., et al., 2010. A comparison of chemical mechanisms based on TRAMP-2006 field data. Atmos. Environ. 44, 4116−4125.

Courant, R., Friedrichs, K., Lewy, H., 1928. Über die partiellen differenzengleichungen der mathematischen physik. Math. Ann. 100, 32−74.

Hertel, O., Berkowicz, R., Christensen, J., 1993. Test of two numerical schemes for use in atmospheric transport-chemistry models. Atmos. Environ. 27A, 2591−2611.

Jacobson, M.Z., Turco, R.P., 1994. SMVGEAR: a sparse-matrix, vectorized Gear code for atmospheric models. Atmos. Environ. 28A, 273−284.

Jet Propulsion Laboratory (JPL), 2006. Chemical kinetics and photochemical data for use in atmospheric studies, evaluation number 15. NASA JPL, Publication 06−2, Pasadena, CA.

Jet Propulsion Laboratory (JPL), 2015. Chemical kinetics and photochemical data for use in atmospheric studies, evaluation number 18. NASA JPL, Publication 15−10, Pasadena, CA.

Lefer, B., Rappenglück, B., Flynn, J., Haman, C., 2010. Photochemical and meteorological relationships during the TexAQS-II Radical and Aerosol Measurement Project (TRAMP). Atmos. Environ. 44, 4005−4013.

Madronich, S., Flocke, S., 1998. The role of solar radiation in atmospheric chemistry. In: Boule, P. (Ed.), Handbook of Environmental Chemistry. Springer-Verlag, Heidelberg, pp. 1−26.

Mao, J., Ren, X., Chen, S., Brune, W.H., Chen, Z., Martinez, M., et al., 2010. Atmospheric oxidation capacity in the summer of Houston 2006: comparison with summer measurements in other metropolitan studies. Atmos. Environ. 44, 4107−4115.

Olaguer, E.P., 2012a. The potential near source ozone impacts of upstream oil and gas industry emissions. J. Air Waste Manage. Assoc. 62, 966−977.

Olaguer, E.P., 2012b. Near source air quality impacts of large olefin flares. J. Air Waste Manage. Assoc. 62, 978−988.

Olaguer, E.P., 2013. Application of an adjoint neighborhood scale chemistry transport model to the attribution of primary formaldehyde at Lynchburg Ferry during TexAQS II. J. Geophys. Res. Atmos. 118, 4936−4946.

Olaguer, E.P., Herndon, S.C., Buzcu Guven, B., Kolb, C.E., Brown, M.J., Cuclis, A.E., 2013. Attribution of primary formaldehyde and sulfur dioxide at Texas City during SHARP/ Formaldehyde and Olefins from Large Industrial Releases (FLAIR) using an adjoint chemistry transport model. J. Geophys. Res. Atmos. 118, 11,317−11,326.

Olaguer, E.P., Lefer, B., 2014. Measurement and Modeling of Ozone Impacts of Oil and Gas Activities in the Uinta Basin. Houston Advanced Research Center, The Woodlands, TX.

Ryerson, T.B., Trainer, M., Angevine, W.M., Brock, C.A., Dissly, R.W., Fehsenfeld, F.C., et al., 2003. Effect of petrochemical industrial emissions of reactive alkenes and NO_x on tropospheric ozone formation in Houston, Texas. J. Geophys. Res.108. Available from: <http:// dx.doi.org/10.1029/2002JD003070>.

Sandu, A., Verwer, J.G., Blom, J.G., Spee, E.J., Carmichael, G.R., Potra, F.A., 1997. Benchmarking stiff ODE solvers for atmospheric chemistry problems II: Rosenbrock solvers. Atmos. Environ. 31, 3459−3472.

Saunders, S.M., Jenkin, M.E., Derwent, R.G., Pilling, M.J., 2003. Protocol for the development of the Master Chemical Mechanism, MCM v3 (Part A): tropospheric degradation of non-aromatic volatile organic compounds. Atmos. Chem. Phys. 3, 161−180.

Wild, O., Zhu, X., Prather, M., 2000. Fast-J: accurate simulation of in- and below-cloud photolysis in tropospheric chemical models. J. Atmos. Chem. 37, 245−282.

Yarwood, G., Whitten, G.Z., Jung, J., Heo, G., Allen, D.T., 2010. Development, evaluation and testing of version 6 of the Carbon Bond Chemical Mechanism (CB6). Final Report to the Texas Commission on Environmental Quality, Austin, TX.

Chapter 11

MultiScale Impact Assessment

Chapter Outline

THE PROBLEM OF MULTIPLE SCALES

Chemistry and transport of air pollution operate on physical scales ranging from very small turbulent eddies to the entire globe. Modeling of atmospheric impacts must somehow recognize process interactions among all these scales, even as they are impossible to explicitly simulate at the current level of computer technology.

Strategies to address the problem of multiple scales in a 3D Eulerian chemical transport model (CTM) are partly determined by the meteorological model that drives the CTM. Atmospheric scales are first separated based on balances among terms in the fundamental fluid equations. For example, on continental and global scales the atmosphere is approximately hydrostatic, that is the vertical pressure gradient is roughly balanced by gravity (Holton, 1979). Global GCMs used in climate simulations assume hydrostatic balance, which filters out atmospheric gravity waves. The effects of gravity waves on momentum, energy, and tracers must then be parameterized along with the effects produced by other sub-grid scale motions. Mesoscale meteorological models assume nonhydrostatic conditions, but their physics breaks down on horizontal scales of roughly 2 km. Micro-scale meteorological models, such as those based on the Large Eddy Simulation (LES) technique, enable smaller-scale atmospheric motions to be credibly simulated based on higher order treatments of turbulence (Wyngaard, 2004).

The horizontal resolution of a CTM is usually limited by that of the driving meteorological model. Sub-grid scale pollutant transport is, as a first resort, parameterized based on a turbulent eddy diffusion coefficient, though this approach does not account for sub-grid scale chemistry. A more sophisticated approach is the Plume-in-Grid (PinG) technique (e.g., Seigneur et al., 1983). Instead of instantly mixing the emissions from sources into an entire grid cell

Atmospheric Impacts of the Oil and Gas Industry. DOI: http://dx.doi.org/10.1016/B978-0-12-801883-5.00011-5
121

volume, a PinG module simulates relevant physical and chemical processes during a sub-grid scale phase, allowing emission plumes to expand and evolve chemically in a realistic manner prior to being merged with the grid cell. A more direct approach is to employ a series of nested grids of increasingly higher resolution and smaller domain size. CTMs with variable resolution or even adaptive grids have also been used to simulate both large-scale and finer features (e.g., Benson and McRae, 1991; Srivastava et al., 2000).

A common modeling approach in NAAQS attainment demonstrations makes use of a regional CTM such as the USEPA's Community Multiscale Air Quality (CMAQ) modeling system (Byun and Ching, 1999) with up to three levels of nested grids, and driven by the Weather Research and Forecasting (WRF) mesoscale meteorological model (Michalakes et al., 2001). Typical resolutions employed for attainment demonstrations are 36 km for the continental scale, 12 km for the regional scale, and 4 km for the urban or fine scale. In succeeding sections, we will consider issues pertaining to modeling demonstrations of ozone attainment in the metropolitan area of San Antonio, Texas, which may be influenced by emissions from oil and gas development in the Eagle Ford Shale (EFS). In particular, we will discuss the results of a coarse-grid regional air quality simulation, as well as those from a micro-scale ozone impact assessment. We will then provide recommendations as to how multiscale modeling assessments of the air quality impacts of the oil and gas industry may be improved in future studies.

OZONE ATTAINMENT IN SAN ANTONIO, TEXAS

As in other shale plays, there has been a dramatic increase in oil and gas production in the EFS over the past several years (see Fig. 11.1). The Alamo Area Council of Governments (AACOG, 2015a) estimates that EFS oil and gas production may increase air emissions of NO_x and VOCs from 121 and 223 tons per day (tpd), respectively in 2012 to 219−423 tpd and 689−1248 tpd, respectively in 2018. The San Antonio region to the north of the EFS risks being designated in nonattainment of the 70 ppb ozone NAAQS, and has in fact exceeded the old 75 ppb standard since 2012 while technically remaining in compliance due to USEPA administrative timelines. Regional modeling by AACOG (2015b) has shown that ozone in the San Antonio area is greatly impacted by ozone transport from other regions in Texas. It is thus important to consider whether or not EFS activity may contribute to ozone nonattainment in San Antonio or even raise rural ozone to levels approaching or exceeding the ozone standard.

Pacsi et al. (2015) conducted simulations with the Comprehensive Air Quality Model with Extensions (CAMx) to analyze the combined ozone impacts of electricity generation in Texas and oil and gas production in the EFS. They found that an increase in NO_x emissions associated with upstream oil and gas production caused a local increase in average daily maximum 8-h

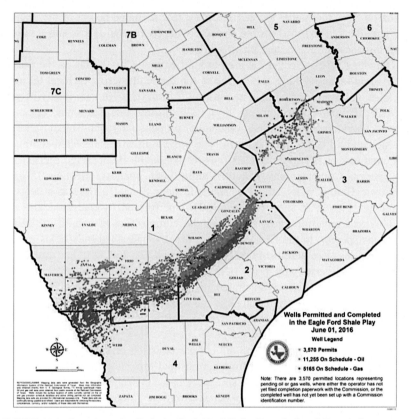

FIGURE 11.1   Locations of wells in the Eagle Ford Shale. *Texas Railroad Commission.*

ozone concentration in South Texas of 0.3−0.7 ppb. However, the CAMx model used by Pacsi et al. (2015), which was at that time largely the same model employed by the State of Texas for ozone attainment demonstrations, had a horizontal resolution of only 12 km in the EFS and South Texas. As demonstrated by Olaguer (2012), this coarse resolution may artificially dilute the highly nonlinear chemistry that occurs in emission plumes. In the following section, we will explore the consequences of using much finer resolution in simulating the ozone impacts of oil and gas activity in the EFS.

A MICRO-SCALE SIMULATION OF AIR QUALITY IN THE EFS

Olaguer (2016) applied the HARC model to the assessment of local ozone impacts in the EFS, as well as possible exports of ozone and precursors towards San Antonio. The HARC model was configured to run over a 12 km × 12 km horizontal domain with horizontal and temporal resolutions of 200 m and 20 s, respectively, as in Olaguer (2012). The vertical resolution increased from the model top at 1 km AGL (the assumed height of the planetary boundary layer) towards the surface, with 10 vertical layers and a minimum thickness of 3.2 m near the ground.

Olaguer (2016) selected the area around Karnes City, Texas as the focus of his assessment (see Fig. 11.2), due to the dense clustering of local oil and gas sites and the presence of a monitoring station (C1070) equipped with an auto-GC. Hourly measurements at C1070 during one particular episode showed elevated concentrations of ethene, propene, NO_x, and other species that signaled a likely emission event. The prevailing wind suggested combustion emissions to the south of C1070 as the source of the observed olefin and NO_x peaks. Coincidentally, satellite night light images reflecting 15-day composite flare activity indicated sporadic operation of a flare at a nearby central production facility. Olaguer (2016) performed inverse modeling of the observed emission event to deduce the composition of natural gas flare emissions specific to the EFS (see Table 11.1), constrained by species ratios from previous flare studies. For example, the HCHO-to-CO molar ratio was set at 5% consistent with prior field measurements (Herndon et al., 2012; Knighton et al., 2012), as illustrated in Fig. 11.3. Because a traffic source could not be ruled out by the inverse model, the synthetic flare composition of Olaguer (2016) should not be viewed as definitive, but merely as the plausible basis for a modeling assessment.

Olaguer (2016) constructed an oil and gas emission inventory for well pads and midstream facilities derived from permits and official databases maintained by Texas state agencies, with flare speciation based on Table 11.1. Mobile emissions for Karnes County were obtained from the Motor Vehicle Emission Simulator (MOVES) model (EPA, 2014) for a 2012 Texas summer weekday simulation. Emissions of NO, NO_2, HONO, and CO generated by MOVES were used to construct approximate link-based hourly emissions of these species for the area of interest by scaling the emissions according to the ratio of the link length in each model grid cell to the total link length in Karnes County. Near-road or large-scale urban ratios of various VOCs to CO from prior observational studies were then employed to specify other species emissions. Biogenic emissions were estimated based on an isoprene emission factor of 150 molC/(km²-h), consistent with aircraft observations by Warneke et al. (2007).

The total OH reactivity of VOCs unresolved by the HARC mechanism was assigned a value of $1.3\,s^{-1}$, mostly reflecting the contribution of alkanes, and based on the analysis of EFS hydrocarbon data by Schade and Roest (2016). Note that the long-lived hazardous air pollutant, benzene (C_6H_6 in Table 11.1), was treated as a passive tracer and included as a transported species in addition to those listed in Table 10.3. Appropriate inflow boundary conditions were derived from regional monitoring stations or from the peer-reviewed literature.

Olaguer (2016) conducted 3-h (11 am−2 pm LST) forward model runs corresponding to both low and high background ozone conditions, with winds from

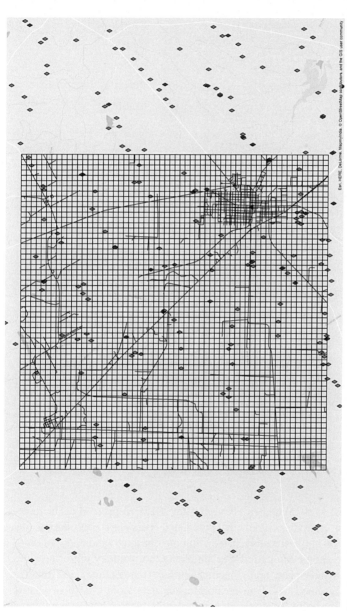

FIGURE 11.2 Illustration of a HARC model domain in the study of Olaguer (2016). Green diamonds indicate well pad locations, while red diamonds indicate midstream facilities. The purple triangle indicates the C1070 monitoring site.

Esri, HERE, DeLorme, MapmyIndia, © OpenStreetMap contributors, and the GIS user community

TABLE 11.1 Flare Composition Inferred From The HARC Inverse Model Olaguer, 2016

Species	Ratio to CO	Ratio to NO$_x$
NO	0.1690	0.9018
NO$_2$	0.0184	0.0982
NO$_x$	0.1874	1.0000
HONO	0.0015	0.0078
HCHO	0.0504	0.2687
CO	1.0000	5.3356
C$_2$H$_4$	0.0733	0.3910
C$_3$H$_6$	0.0241	0.1287
TOL	0.0130	0.0695
XYL	0.0115	0.0612
C$_6$H$_6$	0.0031	0.0166

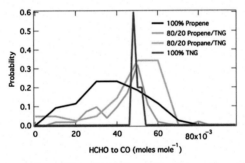

FIGURE 11.3 The molar ratio of formaldehyde to CO for flares with various combinations of Tulsa Natural Gas (TNG), propane and propene. *Scott Herndon, personal communication.*

the south (August 5, 2015) and east (October 22, 2015), respectively. The three-hour simulation period ensured that predicted concentrations during the last hour would be in quasi-steady state, assuming constant wind and emissions during the entire simulation period. For simplicity, horizontally uniform values of background wind speed and direction, surface temperature, and relative humidity were assumed based on meteorological data from C1070 or nearby monitors (see Table 11.2). Vertical extrapolation of the background wind was based on a logarithmic wind profile with a roughness length of 0.1 m, and a Monin–Obukhov length of −100 m. The surface temperature was vertically extrapolated assuming an unstable lapse rate of 12K/km.

TABLE 11.2 Meteorological Parameters in The Simulation of Olaguer (2016)

Time Period	Resultant Wind Speed (m/s)	Resultant Wind Direction (°)	Surface Temperature (K)	Relative Humidity (%)
8/5/15, 11 am−2 pm LST	2.2	177	307.9	55
10/22/14, 11 am−2 pm LST	3.2	102	300.7	54

The major findings of Olaguer (2016) included the following:

- The difference between peak ozone and mean ozone in the 12 km × 12 km modeling domain (the size of a single grid box in the model of Pacsi et al., 2015) was always greater than 4 ppb, indicating a significant contrast in ozone impacts due to local nonlinear chemistry.
- In the mean, the net local effect of emission sources was to decrease ozone relative to the inflow boundary condition due to the titration reaction: $NO + O_3 \rightarrow NO_2 + O_2$, and inhibition of ozone by the conversion of NO_2 to nitric acid. This effect might be mitigated in a regional air quality model by artificial dilution of fresh NO_x plumes.
- Figs. 11.4 and 11.5 illustrate the results of the HARC model simulation for the case of low background ozone (24 ppb) and the model domain of Fig. 11.2. When the wind was from the south, bringing clean Gulf Coast air, primary formaldehyde from combustion sources and co-located emissions of olefins and other reactive VOCs resulted in an increase of up to ~4 ppb in peak ambient ozone within the study area. This ozone production is aided by greater spacing between oil and gas sites in the north−south direction than in the east−west direction, which tilts the balance between titration and inhibition by fresh NO_x sources and photochemical aging towards the latter process.
- Odd oxygen ($O_x = O_3 + NO_2$) at the downwind edge of the model grid is an indication of potential ozone impacts beyond the simulated domain, since NO_2 plumes will eventually be converted to ozone by photolysis further downwind. Regular emissions within the model domain enhanced peak O_x at the downwind edge (toward San Antonio) by ~6 ppb (relative to a background of 24.2 ppb) during southerly flow (see Fig. 11.4).
- Figs. 11.6 and 11.7 illustrate the results of the HARC model simulation for the case of high background ozone (60 ppb) and the model domain of Fig. 11.2. When the wind was from the east, bringing elevated continental background ozone, the east−west orientation of drilling sites resulted in much stronger titration and inhibition than when the wind was from

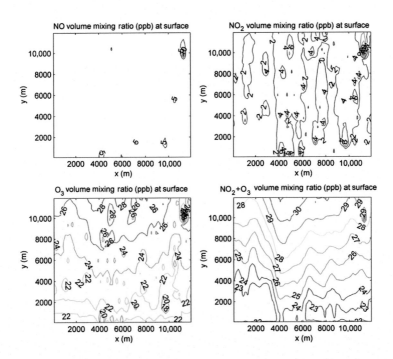

FIGURE 11.4 Final concentrations of NO$_x$ and ozone for southerly flow and low background ozone on August 5, 2015.

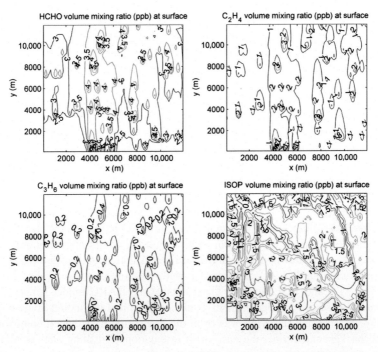

FIGURE 11.5 Final concentrations of reactive VOCs for southerly flow and low background ozone on August 5, 2015.

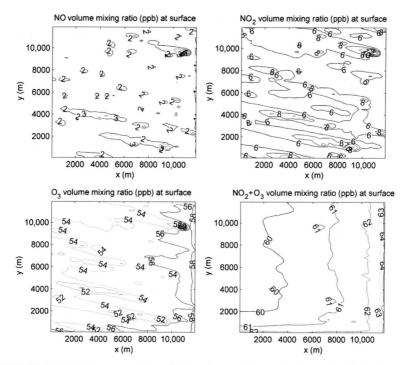

FIGURE 11.6   Final concentrations of NO_x and ozone for easterly flow and high background ozone on October 22, 2015.

the south. Ozone formation was thus locally suppressed. O_3 increases could appear further downwind, especially if air trajectories eventually turn northward.

IMPLICATIONS FOR SIP MODELING

The results of Olaguer (2012, 2016) make it clear that micro-scale approaches deserve some consideration in State Implementation Plan (SIP) modeling. Adaptive grids are not yet used in regulatory applications, partly because the inputs to such models would require extensive development efforts. Because the brute force nesting of micro-scale grids within coarser regional grids may be too computationally expensive, we propose instead the following protocol for the adjustment of coarse-grid regional air quality models used for ozone attainment demonstrations in areas with significant oil and gas development:

1. In the context of specific historical ozone episodes, use regional source apportionment tools to identify those areas within nearby oil and gas basins that are most likely to contribute to ozone nonattainment. Source apportionment tools used in regional modeling applications are described in Appendix B.

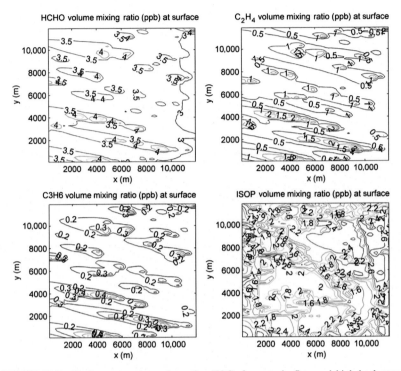

FIGURE 11.7 Final concentrations of reactive VOCs for easterly flow and high background ozone on October 22, 2015.

2. Based on the regional source apportionment analysis, identify serially adjacent coarse grid cells in the attainment demonstration model that correspond to locations with significant oil and gas activities, and which lie on transport trajectories associated with high ozone episodes.

3. For each of the identified coarse grid cells, perform micro-scale modeling similar to that conducted by Olaguer (2016). It may be necessary to expand the treatment of organic nitrate chemistry in the micro-scale model to ensure more accurate treatment of reactive nitrogen reservoirs and enhance the validity of longer simulations.

4. Develop a micro-scale plume correction algorithm based on the predictions of both coarse-grid and micro-scale models. For example, the difference between mean cell ozone predicted by the coarse-grid model and that predicted by the micro-scale model for the same area can be simulated as an ozone source/sink term in the regional model.

5. The spatial contrast patterns inferred through micro-scale modeling can be used to develop a simple geospatial parameterization to determine the magnitude and location of peak ozone plumes within coarse grid cells based on local meteorological parameters, upwind chemical parameters,

and pattern recognition methods. One possible technique that can be used for this purpose is Principal Component Analysis (PCA), as has been applied in other fields (e.g., Vo and Durlofsky, 2014).

The protocol suggested above does not require full-fledged integration between micro-scale and regional air quality models, but rather uses micro-scale models in an off-line mode to develop a simple and cost-effective way to correct the defects of regional models associated with high resolution non-linear chemistry.

REFERENCES

Alamo Area Council of Governments (AACOG), 2015a. Oil and gas emission inventory update. Eagle Ford Shale, San Antonio, Texas.

Alamo Area Council of Governments (AACOG), 2015b. Ozone Analysis, June 2006 Photochemical Modeling Episode. San Antonio, Texas.

Benson RA, McRae DS, 1991. A solution adaptive mesh algorithm for dynamic/static refinement of two and three-dimensional grids. In: Proceedings of the Third International Conference on Numerical Grid Generation in Computational Field Simulations. Barcelona, Spain, June 3−7 (A92-47035 19-64). Amsterdam and New York, North-Holland, pp. 185−199.

Byun, D.W., Ching, J.K.S., 1999. Science algorithms of the EPA Models-3 Community Multiscale Air Quality (CMAQ) modeling system. EPA/600/R-99/030. U.S. EPA, Office of Research and Development.

Environmental Protection Agency (EPA), 2014. MOVES2014a User Guide. EPA-420-B-15-095. Office of Transportation and Air Quality, Washington, DC.

Herndon, S.C., Nelson, D.D., Wood, E.C., Knighton, W.B., Kolb, C.E., Kodesh, Z., et al., 2012. Application of the carbon balance method to flare emissions characteristics. Ind. Eng. Chem. Res. 51, 12577−12585.

Holton, J.R., 1979. An Introduction to Dynamic Meteorology. Academic Press, New York, 391 pp.

Knighton, W.B., Herndon, S.C., Franklin, J.F., Wood, E.C., Wormhoudt, J., Brooks, W., et al., 2012. Direct measurement of volatile organic compound emissions from industrial flares using real-time online techniques: Proton Transfer Reaction Mass Spectrometry and Tunable Infrared Laser Differential Absorption Spectroscopy. Ind. Eng. Chem. Res. 51, 12674−12684.

Michalakes, J., Chen, S., Dudhia, J., Hart, L., Klemp, J., Middlecoff, J., et al., 2001. Development of a next generation regional weather research and forecast model. In: Zwieflhofer, W., Kreitz, N. (Eds.), *Developments in Teracomputing: Proceedings of the Ninth ECMWF Workshop on the Use of High Performance Computing in Meteorology.* World Scientific, Singapore, pp. 269−276.

Olaguer, E.P., 2012. The potential near source ozone impacts of upstream oil and gas industry emissions. J. Air Waste Manage. Assoc. 62, 966−977.

Olaguer EP, 2016. Near source ozone impacts of oil and gas sites in the Eagle Ford Shale. Final report to the Environmental Defense Fund, Austin, TX.

Pacsi, A.P., Kimura, Y., McGaughey, G., McDonald-Buller, E.C., Allen, D.T., 2015. Regional ozone impacts of increased natural gas use in the Texas power sector and development in the Eagle Ford Shale. Environ. Sci. Technol. 49, 3966−3973.

Schade, G.W., Roest, G., 2016. Analysis of non-methane hydrocarbon data from a monitoring station affected by oil and gas development in the Eagle Ford shale, Texas. Elem Sci Anthrop. 4. Available from: http://dx.doi.org/10.12952/journal.elementa.000096.

Seigneur, C., Tesche, T.W., Roth, P.M., Liu, M.K., 1983. On the treatment of point source emissions in urban air quality. Atmos. Environ. 17, 1655–1676.

Srivastava, R.K., McRae, D.S., Odman, M.T., 2000. An adaptive grid algorithm for air-quality modeling. J. Comput. Phys. 165, 437–472.

Vo, H.X., Durlofsky, L.J., 2014. A new differentiable parameterization based on principal component analysis for the low-dimensional representation of complex geological models. Math. Geosci. 46, 775–813.

Warneke, C., McKeen, S., de Gouw, J.A., Del Negro, L., Brioude, J., et al., 2007. Determination of biogenic emissions from aircraft measurements during TEXAQS2006 and ICARTT2004 campaigns and comparison with biogenic emission inventories. AGU Fall Meeting Abst. 1, 5.

Wyngaard, J.C., 2004. Changing the face of small-scale meteorology. In: Fedorovich, E., Rotunno, R., Stevens, B. (Eds.), Atmospheric Turbulence and Mesoscale Meteorology. Cambridge University Press, Cambridge, UK, pp. 17–34.

Chapter 12

Emission Controls

Chapter Outline

OVERVIEW OF MITIGATION STRATEGIES

Mitigating the atmospheric impacts of the oil and gas industry depends ultimately on the ability to control emissions to air from a variety of industrial processes. Here, we provide an overview of mitigation strategies based on different process types.

Fugitive Emissions

Leak detection and repair (LDAR) is one of the most common strategies for the control of fugitive emissions from pipes, valves, flanges, and other types of equipment (EPA, 2014). LDAR requires sufficiently frequent inspection of facilities, a leak detection method such as an OGI camera, and a record-keeping system to keep track of facility component counts, the time needed to survey components, and any repair and maintenance activities.

An example of a common leak problem arises from dissolved gas in crude oil and liquid condensate storage tanks. Pressure vacuum relief valves (PVRVs) used to maintain safe pressure in these tanks are usually unmonitored, which prevents immediate attention to fugitive releases when the tank pressure exceeds a set point. Integrating a wireless transmitter with a PVRV enables rapid detection and mitigation of these releases (Emerson Process Management, 2016).

Vapor recovery units (VRUs) collect and compress the gas built up in liquid petroleum storage tanks, which can then be re-directed to a sales line, used on-site for fuel, or flared rather than vented to the atmosphere (EPA,

Atmospheric Impacts of the Oil and Gas Industry. DOI: http://dx.doi.org/10.1016/B978-0-12-801883-5.00012-7

2006a). An alternative method of controlling tank fugitives is VOC suppression technology, which controls evaporation from hydrocarbon-wetted surfaces by chemical barriers such as foams, surfactant monolayers or nano-droplet aerosol (Unrau, 2016).

Fugitives from compressors require different strategies for emissions control depending on the compressor type (ICF International, 2014). In the case of reciprocating compressors, rod packing systems maintain a seal around the piston rod, minimizing the leakage of high pressure gas from the compressor cylinder. However, the rod packing wears out over time and must be replaced periodically to prevent gas from escaping. Centrifugal compressors with wet seals use circulating oil as a seal against the escape of high pressure gas. The seal oil must be degassed for proper operation of the seal, and the gas removed must be captured to prevent being vented directly to the atmosphere. An alternative is to replace high emission wet seals with lower emitting dry seals, which operate mechanically using hydrodynamic grooves and static pressure.

Blowdown emissions from compressor units can be significantly reduced by keeping systems fully or partially pressurized during shutdown (EPA, 2006b). In addition, blowdown vent lines can be connected to the fuel gas system to recover vented gas. An ejector can also be installed on blowdown vent lines so that the discharge of an adjacent compressor triggers the pumping of blowdown or leaked gas into the suction of an operating compressor or a fuel gas system.

An ejector is an example of a pneumatic controller, which uses the pressure of the natural gas stream to adjust valves, actuate liquid level and temperature controllers, and perform other control functions. These devices are classified as low bleed (emitting < 6 scf/h), high bleed (emitting ≥ 6 scf/h) or intermittent (emitting only when actuating) controllers. Mitigation of leaks from pneumatic controllers usually consists of replacing high bleed and high-emitting intermittent controllers with low bleed controllers (ICF International, 2014).

Pneumatic pumps use gas pressure to drive a fluid by means of a piston, rotating impellers, or a positive displacement (i.e., by trapping a fixed amount of fluid and forcing it into a discharge pipe). Chemical injection pumps inject methanol or other chemicals into gas wells to prevent freezing during cold weather, while larger pneumatic pumps are used to circulate glycol in gas dehydrators. A mitigation measure to prevent these pumps from venting gas when they operate is to replace gas-driven pumps with electric pumps driven by solar energy or local electricity (ICF International, 2014).

Glycol reboilers vent gases to the atmosphere while regenerating glycol for use in dehydration. The amount of venting is directly proportional to the glycol circulation rate, which is often two or three times higher than necessary. Reducing circulation rates decreases emissions at negligible cost, while installing flash tank separators on glycol dehydrators limits emissions even further (EPA, 2006c). Recovered gas can be recycled and/or used as a fuel.

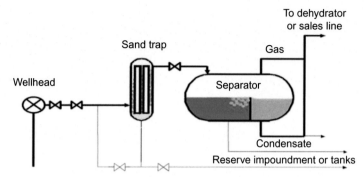

FIGURE 12.1 Reduced Emissions Completion (REC) equipment layout. *USEPA.*

Well completion initiates the flow of petroleum from a newly drilled well prior to production. In a conventional well completion, the flowback from the well is released to an open pit or tank, where the gas released from the liquids is vented to the atmosphere or flared. Green completion or Reduced Emissions Completion (REC) is an alternate practice that reduces venting and flaring of gas (EPA, 2011a). It uses portable equipment to capture and separate produced gas from the liquids and solids in the flowback stream (see Fig. 12.1). The gas stream produced from green completions is then ready or nearly ready for the sales pipeline.

Liquids unloading is the process of removing liquids from the bottom of older, vertical wells whose pressure has declined. One method of liquids unloading uses the pressurized gas in the reservoir to vent the liquids to the atmosphere. An alternative to this method is to use plunger lift systems that fit into the well bore and bring liquids to the surface more efficiently while reducing the amount of venting (ICF International, 2014).

Natural gas pipeline systems require routine maintenance and repair, during which line pressure is reduced and gas discharged from pipeline sections to ensure safe working conditions. Inline and portable compressors can be deployed to lower pipeline pressure to reduce venting during maintenance activities (EPA, 2006d).

Defects such as corrosion, dents, gouges, pits, and cracks can eventually cause pipelines to rupture. Composite wrap can be used to repair nonleaking defects on an operating pipeline without taking it out of service (EPA, 2006e). For gas mains, flexible plastic liners may be inserted in cast iron and unprotected steel pipes to eliminate fugitive emissions (EPA, 2011c).

Combustion Emissions

Stationary internal combustion (IC) engines used by the oil and gas industry can be spark-ignited, gas-fired engines or compression-ignited, diesel-fired engines. A rich-burn engine operates with excess fuel in the combustion chamber, while a lean-burn engine operates with excess air during combustion. Diesel engines inherently operate lean, whereas natural gas-fired engines can be operated in rich or lean modes, depending on the air-to-fuel ratio. A control option for gas-fired engines is to install automatic air-to-fuel ratio controllers to optimize engine operation (EPA, 2011b).

Emission control technologies for stationary IC engines include the use of catalysts (MECA, 2015). A catalytic emission control system consists of a steel housing that contains a metal or ceramic substrate. Interior surfaces on the substrate are coated with catalytic metals, such as platinum (Pt), rhodium (Rh), palladium (Pd), or vanadium (V). These catalysts transform pollutants into less harmful gases by causing chemical reactions in the engine exhaust stream. Oxidation catalysts are used to control CO, VOCs and HAPs from lean-burn gas engines, whereas nonselective catalytic reduction (NSCR) is used to control emissions of NO_x, CO, hydrocarbons and HAPs from rich-burn gas engines. Selective catalytic reduction (SCR) uses ammonia or urea to reduce NO_x emissions from both diesel and lean-burn gas engines, while lean NO_x catalysts (LNCs) accomplish the same by injecting a small amount of diesel fuel or other hydrocarbon reductant into the exhaust upstream of a catalyst.

Diesel oxidation catalysts (DOCs) can be used to control PM as well as gaseous emissions from diesel engines. Diesel particulate filter (DPF) systems use semi-permeable walls to trap soot, which is then converted to CO_2 by a thin layer of catalyst.

An increasingly viable option for controlling combustion emissions from oil and gas sites is flare minimization (HARC, 2015). Methods to reduce flaring include the generation of onsite power using stranded natural gas, waste heat from natural gas, or field gas converted to either compressed or liquefied natural gas (CNG or LNG). Local co-ops might also be recruited to buy back electricity or to gather isolated gas and process it into more transportable CNG or LNG.

BENEFITS OF EMISSIONS CONTROL

The costs and benefits of controlling emissions from the oil and gas industry may be illustrated by using methane as the basis of an economic assessment. ICF International (2014) analyzed strategies for methane mitigation in natural gas systems (see Fig. 12.2). The main conclusions from this assessment were as follows:

- Methane emissions from oil and gas activities were projected to grow by 4.5% from 2011 to 2018 despite reductions expected from the implementation of the USEPA's New Source Performance Standards (NSPS, Subpart OOOO) adopted in 2012. These standards were intended to curb emissions of methane, VOCs, and HAPs from new, reconstructed and modified oil and gas sources.

The natural gas production industry

Natural gas systems encompass wells, gas gathering and processing facilities, storage, and transmission and distribution pipelines.

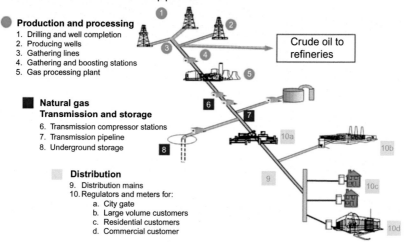

FIGURE 12.2 Natural gas systems for which emission controls may be applied to mitigate methane releases. *USEPA.*

- The oil sector accounted for all of the projected net growth, mainly from flaring and venting of associated gas. Growth from new natural gas sources was expected to be offset by the NSPS and other emission reduction activities.
- Sources in existence in 2011 account for nearly 90% of the projected emissions in 2018. Twenty-two out of over one hundred emission source categories accounted for over 80% of the 2018 emissions.
- A 40% reduction in on-shore methane emissions is achievable with existing technologies at a net total cost of $0.66 per million cubic feet (MCF) of methane reduced, or less than $0.01/MCF of gas produced (see Table 12.1). This would save the US economy over $100 million/year, assuming an initial capital cost of about $2.2 billion.
- Reducing methane emissions would also result in a 44% reduction in VOC and HAP emissions at no extra cost.

UPDATED NSPS

In 2016, the USEPA updated its NSPS for the oil and gas industry (81 FR 35823). The new NSPS rule is expected to reduce methane by 510,000 short tons in 2025, equivalent to 11 million metric tons of CO_2. Climate benefits from the rule are estimated at around $690 million in 2025, compared to estimated costs of $530 million. The rule is also expected to reduce VOCs

TABLE 12.1 Estimated Methane Reductions and Associated Costs/Savings From Various Mitigation Technologies for the Natural Gas Supply Chain

Source/Measure	Annualized Cost ($million/year)	BCF Methane Reduced/year	$/MCF Methane Reduced	Initial Capital Cost ($ million)
Convert gas-driven glycol pumps to electric pumps	−23.4	5.8	−4.05	23.1
Centrifugal compressors (wet seals) gas capture	−58.7	19.1	−3.07	79.6
Compressor stations (storage) leak detection and repair	−4.5	1.5	−3.03	2.8
High bleed pneumatic devices to low bleed	−67.4	25.4	−2.65	246.8
Reciprocating compressor fugitives leak detection and repair	−10.5	32.3	−0.33	61.6
Vapor recovery for condensate tanks without control devices	0.1	0.4	0.21	8.5
Flaring of stranded gas from oil wells instead of venting	2.4	8.2	0.30	228.3
Vapor recovery for oil tanks without control devices	1.8	5.5	0.33	105.1
Pump-down instead of pipeline venting during maintenance	2.3	4.2	0.53	0.0
Use solar power for chemical injection pumps	2.7	4.8	0.57	432.0
Use plunger lifts in place of uncontrolled liquids unloading	1.2	1.6	0.74	27.8
Capture gas from transmission stations instead of venting	7.5	5.9	1.27	49.4

(*Continued*)

TABLE 12.1 (Continued)

Source/Measure	Annualized Cost ($million/ year)	BCF Methane Reduced/ year	$/MCF Methane Reduced	Initial Capital Cost ($ million)
Gathering and boosting stations leak detection and repair	5.0	3.3	1.51	17.7
Intermittent bleed pneumatic devices to low bleed	20.9	12.1	1.72	455.4
Flare gas from oil well completions with hydraulic fracturing	14.5	6.8	2.13	50.4
Compressor stations (storage) leak detection and repair	7.7	2.8	2.79	5.3
Well fugitives leak detection and repair	43.9	12.5	3.51	84.4
Replace worn out rod packing on reciprocating compressors	22.3	3.6	6.11	182.3
Local distribution meters and regulators leak detection and repair	140.6	7.1	19.75	91.5
Grand Total	**108.3**	**162.9**	**0.66**	**2,151.9**

Source: ICF International, 2014. Economic analysis of methane emission reduction opportunities in the US onshore oil and natural gas industries. Report to the Environmental Defense Fund, Austin, TX. Available at: < https://www.edf.org/energy/icf-methanecost-curve-report > (accessed 30.06.16).

by 210,000 tons and HAPs by 3900 tons in 2025. Various aspects of the updated NSPS are summarized below.

Leak Detection and Repair

Owners or operators of oil and gas facilities must find and repair leaks according to the following provisions:

- A fixed schedule must be used for monitoring leaks rather than a schedule that varies with performance. For well sites, leaks must be monitored twice a year, while compressor stations must conduct quarterly leaks monitoring.

- Either OGI cameras or portable analyzers suitable for implementing EPA Method 21 may be used to find and repair leaks.
- Emerging, innovative technologies may also be used to monitor leaks with EPA approval.

Pneumatic Pumps and Controllers

At natural gas well sites, methane and VOC emissions must be routed from pneumatic pumps used to transfer fluids or to circulate glycol to an on-site control device or process. Owners or operators of natural gas well sites are encouraged to adopt solar-powered, electrically-powered or air-driven pumps in place of pumps driven by natural gas. In the case of gas processing plants, the use of zero-emissions pneumatic pumps is mandatory. High bleed continuous pneumatic controllers at natural gas transmission stations must be replaced with low bleed controllers.

Compressors

Fugitive emissions from compressors at natural gas transmission stations must be controlled. Methane and VOC emissions from centrifugal compressors with wet seal systems must be reduced by 95% through flaring, or by routing captured gas back to a process. Owners or operators are encouraged to use centrifugal compressors with dry seal systems where feasible. Rod packing systems in reciprocating compressors must be replaced at least every 26,000 hours of operation or every 36 calendar months. Operators may, as an alternative, route emissions from the rod packing via a closed vent system under negative pressure to another process or piece of equipment.

Green Completions

Owners or operators of hydraulically fractured oil wells must use green completion methods to capture emissions. Exceptions to this rule include new exploratory ("wildcat") wells, delineation wells (used to define the borders of a reservoir), low-pressure wells, or wells without access to a pipeline. While green completions are not required for these wells, emissions from completion processes must still be controlled using combustion.

REFERENCES

Environmental Protection Agency (EPA), 2006a. Installing vapor recovery units on storage tanks. Lessons Learned from Natural Gas STAR Partners. Available at: < https://www3. epa.gov/gasstar/documents/ll_final_vap.pdf > (accessed 01.07.16).

Environmental Protection Agency (EPA), 2006b. Reducing emissions when taking compressors off-line. Lessons Learned from Natural Gas STAR Partners. Available at: < https://www3. epa.gov/gasstar/documents/ll_compressorsoffline.pdf > (accessed 02.07.16).

Environmental Protection Agency (EPA), 2006c. Optimize glycol circulation and install flash tank separators in glycol dehydrators. Lessons Learned from Natural Gas STAR Partners. Available at: < https://www3.epa.gov/gasstar/documents/ll_flashtanks3.pdf > (accessed 02.07.16).

Environmental Protection Agency (EPA), 2006d. Using pipeline pump-down techniques to lower gas line pressure before maintenance. Lessons Learned from Natural Gas STAR Partners. Available at: < https://www3.epa.gov/gasstar/documents/ll_pipeline.pdf > (accessed 02.07.16).

Environmental Protection Agency (EPA), 2006e. Composite wrap for non-leaking pipeline defects. Lessons Learned from Natural Gas STAR Partners. Available at: < https://www3. epa.gov/gasstar/documents/ll_compwrap.pdf > (accessed 02.07.16).

Environmental Protection Agency (EPA), 2011a. Reduced emissions completions for hydrauli-cally fractured natural gas wells. Lessons Learned from Natural Gas STAR Partners. Available at: < https://www3.epa.gov/gasstar/documents/reduced_emissions_completions. pdf > (accessed 01.07.16).

Environmental Protection Agency (EPA), 2011b. Install automated air/fuel ratio controls. Partner Reported Opportunities (PROs) for Reducing Methane Emissions. PRO fact sheet number 104. Available at: < https://www3.epa.gov/gasstar/documents/auto-air-fuel-ratio.pdf > (accessed 02.07.16).

Environmental Protection Agency (EPA), 2011c. Insert gas main flexible liners. Partner Reported Opportunities (PROs) for Reducing Methane Emissions. PRO fact sheet number 402. Available at: < https://www3.epa.gov/gasstar/documents/insertgasmainflexibleliners. pdf > (accessed 02.07.16).

Environmental Protection Agency (EPA), 2014. Leak Detection and Repair: A Best Practices Guide. Office of Enforcement and Compliance Assurance, Washington, DC. Available at: < https://www.epa.gov/sites/production/files/2014-02/documents/ldarguide.pdf > (accessed 01.07.16).

Emerson Process Management, 2016. Benefits of wireless monitoring of tank storage pressure safety valves. White paper D352445X012. McKinney, TX.

Houston Advanced Research Center (HARC), 2015. Recommendations to address flaring issues, solutions and technologies. White paper, Environmentally Friendly Drilling Program. Available at: < http://efdsystems.org/pdf/FIST_Whitepaper_May_2015_Updated_rev.pdf > (accessed 01.07.16).

ICF International, 2014. Economic analysis of methane emission reduction opportunities in the U.S. onshore oil and natural gas industries. Report to the Environmental Defense Fund, Austin, TX. Available at: < https://www.edf.org/energy/icf-methanecost-curve-report > (accessed 30.06.16).

Manufacturers of Emission Controls Association (MECA), 2015. Emission control technology for stationary internal combustion engines. Arlington, VA. Available at: < http://www.meca.org/ resources/MECA_stationary_IC_engine_report_0515_final.pdf > (accessed 02.07.16).

Unrau, C.J., 2016. VOC suppression technology. Presented at the Houston Advanced Research Center, May 10, The Woodlands, TX.

Epilogue: The Road Ahead

Deteriorating air quality and accelerated radiative forcing of climate are increasingly associated with oil and gas development. Nevertheless, there are significant means to address these concerns. We have seen that a number of advanced technologies are already available to identify and mitigate causes of air pollution from the oil and gas industry. A variety of real time monitoring methods can now detect emissions from individual process units while operating inside or outside facility fence lines, with or without industry cooperation. These emissions can even be quantified in near real time using data assimilation and inverse modeling techniques implemented on fast computer workstations or Cloud supercomputers, based on immediate broadcasts of ambient measurements over the Internet. Such information can help operators of industrial facilities rapidly fix problems and apply the latest control measures, thereby saving valuable product from wasteful venting to the atmosphere. The general public may now also be alerted to the occurrence of large emission events, so that needless exposure to hazardous air pollutants can be avoided.

Regulators should also be able to keep better track of industrial releases via emission inventories that can be routinely verified by new technologies. This would improve the enforcement of regulations, and enhance the design of emissions trading schemes, flexible permit programs, or other incentive-based measures to limit air pollution. Accurate emission inventories, together with more sophisticated and comprehensive (possibly adaptive grid) multi-scale models, will in addition enable policymakers to better anticipate the environmental impacts of mitigation strategies, and to compare and evaluate different control options.

All these improvements are not only possible, but feasible. Human nature, however, does not always respond in optimal ways to opportunity. Fear of change and of threats to the *status quo* must inevitably be faced in order to ensure a sustainable civilization. The question that decision makers need to ask is, "What prevents us from taking the road ahead?"

Appendix A

Conventional Air Quality Monitoring Stations

TYPES OF MONITORING STATIONS

The US Clean Air Act requires each state to establish a network of air monitoring stations for criteria pollutants based on location and operational norms set by the EPA, which refers to these sites as State and Local Air Monitoring Stations (SLAMS). While SLAMS are mainly intended to monitor criteria pollutants, some stations also provide local measurements of HAPs. States must provide the EPA with an annual summary of measurements from each SLAMS monitor, with more detailed data to be made available upon request.

According to the EPA (1998a), the objectives of SLAMS are to determine:

- the highest pollutant concentrations expected to occur in the monitored region;
- representative concentrations in areas of high population density;
- the impact on ambient pollution levels of significant sources or source categories;
- general background concentration levels;
- the extent of regional pollutant transport; and
- impacts affecting welfare in more rural and remote areas (such as visibility impairment and effects on vegetation).

A subset of SLAMS sites is known as Photochemical Assessment Monitoring Stations (PAMS). These stations must measure O_3, NO_x, and surface meteorological parameters (wind speed and direction, ambient temperature, barometric pressure, relative humidity, precipitation, and solar radiation) on an hourly basis. Most PAMS sites measure 56 target hydrocarbons every 1 to 3 hours, while some sites also collect data on 3 carbonyl compounds (formaldehyde, acetaldehyde, and acetone) every 3 hours during the ozone season. The objective of the PAMS network is to provide data that enable: (1) the development of effective ozone control strategies; (2) evaluation and improvement of photochemical grid models; (3) reconciliation of emissions inventories; (4) characterization of ozone, ozone precursor, and meteorological trends; (5) determination of ozone attainment; and (6) assessment of population exposure.

PAMS MEASUREMENT TECHNIQUES

The US EPA has defined a federal reference method (FRM) for the measurement of each criteria pollutant. Only measurement techniques defined in the various appendices in the US Code of Federal Regulations, Title 40, Part 50 can be considered FRMs. However, the EPA also allows the use of equivalent methods for air quality monitoring. A federal equivalent method (FEM) is any measurement technique that has been acknowledged by the EPA as yielding equivalent results to the FRM based on rigorous field tests. In this appendix, we describe the various techniques used at PAMS sites to measure ozone and its precursors, including some HAPs such as BTEX, formaldehyde, and acetaldehyde (see EPA, 1998b).

The FRM for O_3 measurement is based on the detection of chemiluminescence resulting from the reaction of O_3 with ethylene gas. However, most SLAMS use UV absorption photometry as a FEM. This technique is based on the absorption by ozone of light from a mercury lamp at a wavelength of 254 nm.

NO and NO_2 are typically measured using a chemiluminescence instrument based on the gas-phase reaction of NO and O_3. To measure NO_x, NO_2 is first converted to NO using either molybdenum as a reductant or high energy photons. The concentration of NO_2 is then calculated as the difference between NO_x and NO measured at the same point in time.

Total nonmethane organic compounds (TNMOC) are measured by cryogenically trapping an air sample and then desorbing the trapped material into a flame ionization detector (FID), which counts the ions formed during combustion of organic compounds in a hydrogen flame. Automated measurement of speciated hydrocarbons (including BTEX) in a PAMS station is achieved by introducing desorbed material from a cryogenic trap to a carrier gas and passing the mixture through a gas chromatograph (GC) prior to detection with a FID. The analytical column in the GC then separates the mixture into individual components based on the equilibrium between the mobile (carrier gas) and stationary (liquid column coating) phases. The separated components elute from the column and enter the FID, which generates a signal based on the number of carbon atoms present. The time of elution and detection (retention time) is the basis for identifying individual compounds in the air sample.

Measurement of carbonyls (including formaldehyde and acetaldehyde) involves air sample collection on a cartridge containing 2,4-dinitrophenylhydrazine (DNPH), which derivatizes the carbonyl compounds. Because ozone is an interferent in the measurement technique, it must first be scrubbed from the sample air stream with potassium iodide (KI) prior to analysis. High performance liquid chromatography (HPLC) with UV/visible detection is the standard PAMS method for analyzing carbonyl derivatives. A solvent or mobile phase carries the sample into the HPLC column. The column packing material or stationary phase then separates the mobile phase material into individual dyes that move in separate bands at different speeds. As the separated dye bands leave the column, they pass into a flow cell that detects each separated compound band based on the absorbance of monochromatic light against a reference beam. The magnitude of the absorbance is related to the concentration of analyte passing through the flow cell.

REFERENCES

Environmental Protection Agency (EPA), 1998a. SLAMS/NAMS/PAMS network review guidance. EPA-454/R-98-003. Office of Air Quality Planning and Standards, Research Triangle Park, NC.

Environmental Protection Agency (EPA), 1998b. Technical assistance document for sampling and analysis of ozone precursors. EPA/600-R-98/161. National Exposure Research Laboratory, Research Triangle Park, NC.

Appendix B

Source Apportionment Techniques

THE NATURE OF SOURCE APPORTIONMENT

Source apportionment is the practice of quantifying the amount of pollution that sources or source categories contribute to ambient air concentrations. Three main approaches for source apportionment are: (1) receptor models, (2) source-oriented (forward) models, and (3) inverse models. Receptor models analyze the ambient environment at the point of impact (i.e., the receptor), whereas source-oriented models account for pollutant transport and transformations beginning at the source and ending at the receptor site. Inverse models reverse this process, starting with measurements at the receptor and computing backwards to obtain information about the sources. Inverse models were described in Chapter 9, Data Assimilation and Inverse Modeling. In this appendix, we discuss receptor models and source-oriented models in some detail.

RECEPTOR MODELS

The recent review by Hopke (2016) summarizes receptor modeling methods for source apportionment. A receptor model (RM) apportions the concentrations of a set of atmospheric pollutants measured at a receptor to emission sources by using multivariate analysis to solve a mass balance equation. The most commonly used RM techniques are Chemical Mass Balance (CMB) and Positive Matrix Factorization (PMF).

CMB is useful when the contributing sources are known, and chemical transformations can be ignored so that mass conservation of the emitted species between the source and receptor can be assumed. CMB uses a least squares method to estimate source contributions on the basis of chemical fingerprints of the known sources, and measured concentrations of pollutants at the receptor. In CMB, a mass balance equation accounts for m chemical species in n measurement samples as contributions from p independent sources as follows:

$$x_{ij} = \sum_{k=1}^{p} g_{ik} f_{kj} + e_{ij} \qquad (B.1)$$

where x_{ij} is the concentration of the jth chemical species ($j = 1, \ldots, m$) measured in the ith sample ($i = 1, \ldots, n$), f_{kj} is the concentration of the jth species in material from the kth source ($k = 1, \ldots, p$), that is the chemical fingerprint of the source, g_{ik} is the airborne contribution of material from the kth source to the ith sample, and e_{ij} is the unexplained portion of the observations. The principal output of CMB is the matrix g_{ik}.

PMF is useful, when the nature of the sources is unknown. It is based on uncertainty-weighted factor analysis, and relies on the solution of an eigenvector-eigenvalue problem wherein the source profile f_{kj} must be deduced from the observations as well as g_{ik}. PMF solves the problem by minimizing a weighted objective function:

$$Q = \sum_{i=1}^{n} \sum_{j=1}^{m} \left(\frac{e_{ij}}{s_{ij}}\right)^2 = \sum_{i=1}^{n} \sum_{j=1}^{m} \left(\frac{x_{ij} - \sum_{k=1}^{p} g_{ik} f_{kj}}{s_{ij}}\right)^2 \qquad (B.2)$$

where s_{ij} is an estimate of the uncertainty for the jth species in the ith sample.

Receptor models are often combined with methods that exploit wind information. For example, the conditional probability function (CPF) is defined as:

$$CPF_{\Delta\theta} = \frac{m_{\Delta\theta}}{n_{\Delta\theta}} \qquad (B.3)$$

where $m_{\Delta\theta}$ is the number of pollution event occurrences from wind sector $\Delta\theta$ that exceed a threshold criterion, and $n_{\Delta\theta}$ is the total number of data from the same wind sector (Kim et al., 2003). Sources are likely to be located at directions that have high conditional probability values.

SOURCE-ORIENTED MODELS

Source apportionment using receptor models becomes difficult when reactive chemistry is important, as is the case with regional ozone formation. Source-oriented models that take complex atmospheric chemistry explicitly into account are more useful in this case. A variety of forward models may be deployed as the basis for this type of source apportionment, including Gaussian reactive plume models, Lagrangian trajectory models, and Eulerian grid models. We illustrate the utility of source-oriented models using Ozone Source Apportionment Technology (OSAT) within the CAMx regional air quality model (Dunker et al., 2002) as an example.

OSAT provides information about the relationships between ozone concentrations and sources of precursors, including initial and boundary conditions as well as emissions, so that the source contributions add up to exactly 100% of the total ozone. The emissions contributions can be attributed to specific geographic areas and/or source categories. OSAT determines whether ozone formation is NO_x or VOC limited in each grid cell and time step, and attributes ozone production according to the relative contributions of the limiting precursor from different sources. Both the precursors and their ozone contributions are tracked using a reactive tracer approach, with separate tracers for ozone attributed to NO_x and VOC.

A second example of the source-oriented modeling approach is provided by Zhang et al. (2013), who used source-tagged species in the US EPA's CMAQ model to perform source apportionment of formaldehyde (HCHO) in Southeast Texas for a two-week long episode (August 28−September 12, 2006) during the TexAQS II field study. The model employed a special hourly emission inventory prepared by the Texas Commission on Environmental Quality (TCEQ) for point sources in the Houston-Galveston-Brazoria and Beaumont-Port Arthur areas, supplemented by the 2005 National Emissions Inventory. Zhang et al. (2013) found that throughout most of Southeast Texas, primary HCHO accounted for approximately 20−30% of total HCHO. Biogenic sources, natural gas combustion, and motor vehicles in urban areas of Houston respectively accounted for 10−20%, 10−30%, and 20−60% of total primary HCHO. The model-predicted source contributions to HCHO at the University of Houston measurement site generally agreed with source apportionment results obtained by Buzcu Guven and Olaguer (2011) using the PMF technique.

REFERENCES

Buzcu Guven, B., Olaguer, E.P., 2011. Ambient formaldehyde source attribution in Houston during TexAQS II and TRAMP. Atmos. Environ. 45, 4272−4280.

Dunker, A.M., Yarwood, G., Ortmann, J., Wilson, G.M., 2002. Comparison of source apportionment and source sensitivity of ozone in a three-dimensional air quality model. Environ. Sci. Technol. 36, 2953−2964.

Hopke, P.K., 2016. Review of receptor modeling methods for source apportionment. J. Air Waste Manage. Assoc. 66, 237−259.

Kim, E., Hopke, P.K., Edgerton, E.S., 2003. Source identification of Atlanta aerosol by positive matrix factorization. J. Air Waste Manage. Assoc. 53, 731−739.

Zhang, H., Li, J., Ying, Q., Buzcu Guven, B., Olaguer, E.P., 2013. Source apportionment of formaldehyde during TexAQS 2006 using a source-oriented chemical transport model. J. Geophys. Res. Atmos. 118, 1525−1535.

Index

Note: Page numbers followed by "*f*" and "*t*" refer to figures and tables, respectively.

A

Adjoint method, demonstration of, 97–98
Adjoint transport model, 94–97
Aerosol, secondary, 50–52, 51*f*
Air pollution, long-range transport of, 34, 36
Air toxics, 25
Aliso Canyon gas leak, 23–24
Alleged health effects, of oil and gas site emissions, 26–27
Ambient air monitoring, 79
  advanced measurement technologies, 85–89
    chemical techniques, 89
    ionization techniques, 88–89
    optical techniques, 86–88
  application types, monitoring, 79–80
    baseline and trend monitoring, 80
    mapping and surveying, 79–80
    source attribution and emissions quantification, 80
  sampling strategies, 80–85
    distributed sensor networks, 85
    mobile platforms with real-time sensors, 81
    remote sensing, 81–85
Amine regenerators, 16, 16*f*
Aromatic reactions, HARC model, 109*t*
Associated natural gas, 60–61
Atmospheric ozone, 31–32
  design value (DV), 32
Atmospheric pollution, 27
Atmospheric transformations, 103–104
Atmospheric window, 55–56
Auto-GC stations, 26

B

Barnett Coordinated Campaign, 60
Barnett Shale, 8, 26, 40*f*
  emissions from, 39, 39*t*
  methane emissions in, 60, 61*f*

Baseline and trend monitoring, 80
Benzene, 9, 24, 25*f*, 27
  exposure to, 23–24
Benzene, toluene, ethyl benzene, and xylenes (BTEX), 9, 15, 97, 146
Benzene and other Toxics Exposure (BEE-TEX), 24–25, 25*f*, 28, 84*f*
Blowdown emissions estimated from displacement equation, 72
Blowdown emissions from compressor units, 134
Blowdown events, 17
"Brute force" (BF) sensitivities, 115
BVOC, 104–108

C

C2-benzenes, 98
Carbon dioxide, 7
  emissions, 58, 60–61
  from global gas flaring, 60–61
Carbon monoxide, 31, 62
Carbon tetrafluoride, 55–56
Catalytic emission control system, 136
Cavity Ring Down Spectroscopy (CRDS), 86
Centrifugal compressors, 134
Chemical Ionization Mass Spectrometry (CIMS), 88
Chemical Jacobian terms, 116*t*
Chemical Mass Balance (CMB), 149–150
Chemically reactive emissions, inverse modeling of, 116–118
Chloro-fluorocarbons (CFCs), 32
Christmas tree, 12–13, 13*f*
Climate change, physics of, 55–57
Climate feedback, 57*f*
  positive, 56
Cloud condensation nuclei (CCN), 51
Cloud feedback, 56
CMAQ model, 151
Coarse particulates, 47, 52

ted in the United States
ookmasters